An Introduction
to Lagrangian Mechanics

An Introduction
to Lagrangian Mechanics

Alain J. Brizard

Saint Michael's College, USA

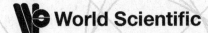

World Scientific

NEW JERSEY · LONDON · SINGAPORE · BEIJING · SHANGHAI · HONG KONG · TAIPEI · CHENNAI

Published by

World Scientific Publishing Co. Pte. Ltd.

5 Toh Tuck Link, Singapore 596224

USA office: 27 Warren Street, Suite 401-402, Hackensack, NJ 07601

UK office: 57 Shelton Street, Covent Garden, London WC2H 9HE

British Library Cataloguing-in-Publication Data
A catalogue record for this book is available from the British Library.

ISBN-13 978-981-281-836-2
ISBN-10 981-281-836-7
ISBN-13 978-981-281-837-9 (pbk)
ISBN-10 981-281-837-5 (pbk)

Printed in Singapore.

To the memory of my father

Yvon Brizard (1929-2007)

Preface

The structure of the present lecture notes on the Lagrangian mechanics of particles and fields is based on achieving several goals. As a first goal, I wanted to model these notes after the wonderful monograph of Landau and Lifschitz on *Mechanics* [12], which is often thought to be too concise for most undergraduate students. One of the many positive characteristics of Landau and Lifschitz's *Mechanics*, however, is that Lagrangian mechanics is introduced in its first chapter and not in later chapters as is usually done in more standard textbooks used at the sophomore/junior undergraduate level.[1] Consequently, the Lagrangian method becomes the centerpiece of the present course and provides a continuous thread throughout the text. This course has been taught at Dartmouth College and Saint Michael's College in approximately the same format proposed in these lecture notes.

As a second goal, the lecture notes introduce several numerical investigations of dynamical equations appearing throughout the text. These numerical investigations present an interactive pedagogical approach, which should enable students to begin their own numerical investigations. As a third goal, an attempt was made to introduce historical facts (whenever appropriate) about the pioneers of Classical Mechanics. Much of the historical information included in the Notes is taken from excellent books by René Dugas [4], Wolfgang Yourgrau and Stanley Mandelstam [18], and Cornelius Lanczos [11]. In fact, from a pedagogical point of view, this historical perspective helps educating undergraduate students in establishing the deep connections between Classical and Quantum Mechanics, which are often ignored or even inverted (as can be observed when students are surprised

---

[1] The reader is invited to read *A call to action* by E. F. Taylor [Am. J. Phys. **71**, 423-425 (2003)], which promotes a reorganization of undergraduate physics education that includes an early introduction of Lagrangian Mechanics (the Principle of Least Action) into the physics curriculum.

to learn that Hamiltonians have an independent classical existence). As a fourth and final goal, I wanted to keep the scope of these notes limited to a one-semester course in contrast to standard textbooks, which often include an extensive review of Newtonian Mechanics as well as additional material such as Hamiltonian chaos.

It is expected that students taking this course will have had a one-year calculus-based introductory physics course followed by a one-semester course in Modern Physics. Ideally, students should have completed their full calculus sequence and, perhaps, have taken a course on ordinary differential equations. On the other hand, this course should be taken before a rigorous course in Quantum Mechanics in order to provide students with a sound historical perspective involving the connection between Classical Physics and Quantum Physics. Hence, the fall semester of the junior year provides a perfect niche for this course. Topics identified with an asterisk can also be included in a more advanced course.

The standard topics covered in these notes are: The Calculus of Variations (Chapter 1), Lagrangian Mechanics (Chapter 2), Hamiltonian Mechanics (Chapter 3), Motion in a Central Field (Chapter 4), Collisions and Scattering Theory (Chapter 5), Motion in a Non-Inertial Frame (Chapter 6), Rigid Body Motion (Chapter 7), Normal-Mode Analysis (Chapter 8), and Continuous Lagrangian Systems (Chapter 9). Each chapter contains a set of problems with variable level of difficulty. Lastly, in order to ensure a self-contained presentation, a summary of mathematical methods associated with linear algebra and numerical analysis is presented in Appendix A. Appendix B presents a brief introduction to the applications of the Jacobi and Weierstrass elliptic functions in Classical Mechanics; see Whittaker's textbook [17] for many more applications. Lastly, Appendix C presents a brief summary of differential geometric methods in the modern formulation of Hamiltonian mechanics and perturbation theory.

Several innovative topics not normally discussed in standard undergraduate textbooks are included throughout the notes. In Chapter 1, a complete discussion of Fermat's Principle of Least Time is presented, from which a generalization of Snell's Law for light refraction through a nonuniform medium is derived and the equations of geometric optics are obtained [3]. We note that Fermat's Principle proves to be an ideal introduction to variational methods in the undergraduate physics curriculum since students are already familiar with Snell's Law of light refraction.

In Chapter 2, we establish the connection between Fermat's Principle of Least Time and Maupertuis-Jacobi's Principle of Least Action. In particular, Jacobi's Principle introduces a geometric representation of single-particle dynamics that establishes a clear pre-relativistic connection between Geometry and Physics. Next, the nature of mechanical forces (e.g., active versus passive forces) is discussed within the context of d'Alembert's Principle, which is based on a dynamical generalization of the Principle of Virtual Work. Lastly, the fundamental link between the energy-momentum conservation laws and the symmetries of the Lagrangian function is first discussed through Noether's Theorem and then Routh's procedure to eliminate ignorable coordinates is applied to a Lagrangian with symmetries.

In Chapter 3, we present a brief discussion of Hamiltonian optics and the wave-particle duality that established the connection between Classical Physics and Quantum Physics. The problem of charged-particle motion in an electromagnetic field is also investigated by the Lagrangian method in the three-dimensional configuration space and the Hamiltonian method in six-dimensional phase space. This important physical example presents a clear link between the Lagrangian and Hamiltonian methods. In Chapter 4, we discuss the role of the Laplace-Runge-Lenz vector invariant in determining the shape of the Kepler bounded orbit. We also use the Laplace-Runge-Lenz vector to study the precession of a perturbed Keplerian orbit. In Chapter 5, we present a complete solution of the soft-sphere scattering problem as well as the problem of elastic scattering by a hard surface. In Chapter 9, we present the variational derivations of the Schroedinger equation and the Euler equations for a perfect fluid. Using the Noether method, we also derive their respective conservation laws.

In Appendix B, we present an introduction to the applications of the Jacobi and Weierstrass elliptic functions in Classical Mechanics. These interesting functions used to be part of the standard curriculum in Classical Mechanics [17, 12] and have now all but disappeared from modern textbooks [7, 13]. For the Jacobi elliptic function, we consider the problems of motion in a quartic potential, while for the Weierstrass elliptic function, we consider the problem of motion in a cubic potential. The problem of the planar pendulum is used to establish the connection between the Jacobi and Weierstrass elliptic functions. Lastly, in Appendix C, we present a brief introduction to noncanonical Hamiltonian mechanics and canonical Hamiltonian perturbation theory.

My interest in Lagrangian Mechanics was awakened more than 30 years ago when I was an undergraduate student at the Collège Militaire Royal de Saint Jean (Canada). One of my professors (Fernand Ledoyen) bravely taught me Lagrangian Mechanics with Landau and Lifschitz [12] and Arnold

[1] as our constant companions. I remember being immediately struck by the beauty of Lagrangian Mechanics and the power of its methods. I have used Lagrangian methods in my own research in plasma physics for the past 20 years. I would like to thank my *Lagrangian* collaborators Allan N. Kaufman (University of California at Berkeley) and Eugene (Gene) R. Tracy (College of William and Mary) for their friendship and support during this time.

Lastly, I owe a great debt of love and gratitude to my wife (Dinah Larsen) and son (Peter Brizard Larsen) and I thank them for their patience and understanding during the arduous process of writing this book.

Alain Jean Brizard

Contents

Chapter 1

The Calculus of Variations

A wide range of equations in physics, from quantum field and superstring theories to general relativity, from fluid dynamics to plasma physics and condensed-matter theory, are derived from action (variational) principles [2, 15]. The purpose of this Chapter is to introduce the methods of the Calculus of Variations that figure prominently in the formulation of action principles in physics.

1.1 Foundations of the Calculus of Variations

1.1.1 *A Simple Minimization Problem*

It is a well-known fact that the shortest distance between two points in Euclidean space is calculated along a straight line joining the two points. Although this fact is intuitively obvious, we begin our discussion of the problem of minimizing certain integrals in mathematics and physics with a search for an explicit proof. In particular, we prove that the straight line $y_0(x) = mx$ yields a path of shortest distance between the two points $(0,0)$ and $(1,m)$ on the (x,y)-plane as follows.

First, we consider the length integral

$$\mathcal{L}[y] = \int_0^1 \sqrt{1 + (y')^2}\, dx, \tag{1.1}$$

where $y' = y'(x)$ and the notation $\mathcal{L}[y]$ is used to denote the fact that the value of the integral (1.1) depends on the choice we make for the function $y(x)$; thus, $\mathcal{L}[y]$ is called a *functional* of y. We insist, however, that the function $y(x)$ satisfy the boundary conditions $y(0) = 0$ and $y(1) = m$. Next, we introduce the modified function

$$y(x; \epsilon) = y_0(x) + \epsilon\, \delta y(x),$$

where $y_0(x) = mx$ and the variation function $\delta y(x)$ is required to satisfy the prescribed boundary conditions $\delta y(0) = 0 = \delta y(1)$. We thus define the modified length integral

$$\mathcal{L}[y_0 + \epsilon\,\delta y] = \int_0^1 \sqrt{1 + (m + \epsilon\,\delta y')^2}\, dx$$

as a function of ϵ and a functional of δy. We now show that the function $y_0(x) = mx$ minimizes the integral (1.1) by evaluating the following derivatives

$$\left(\frac{d}{d\epsilon}\mathcal{L}[y_0 + \epsilon\,\delta y]\right)_{\epsilon=0} = \frac{m}{\sqrt{1 + m^2}} \int_0^1 \delta y'\, dx$$

$$= \frac{m}{\sqrt{1 + m^2}}\left[\delta y(1) - \delta y(0)\right] = 0,$$

and

$$\left(\frac{d^2}{d\epsilon^2}\mathcal{L}[y_0 + \epsilon\,\delta y]\right)_{\epsilon=0} = \int_0^1 \frac{(\delta y')^2}{(1 + m^2)^{3/2}}\, dx > 0,$$

which holds for a fixed value of m and all variations $\delta y(x)$ that satisfy the conditions $\delta y(0) = 0 = \delta y(1)$. Hence, we have shown that $y(x) = mx$ minimizes the length integral (1.1) since the first derivative (with respect to ϵ) vanishes at $\epsilon = 0$, while its second derivative is positive at $\epsilon = 0$. We note, however, that our task was made easier by our knowledge of the actual minimizing function $y_0(x) = mx$; without this knowledge, we would be required to choose a trial function $y_0(x)$ and test for all variations $\delta y(x)$ that vanish at the integration boundaries.

Another way to tackle this minimization problem is to find a way to characterize the function $y_0(x)$ that minimizes the length integral (1.1), for *all* variations $\delta y(x)$, without actually solving for $y(x)$. For example, the characteristic property of a straight line $y(x)$ is that its second derivative vanishes for all values of x. The methods of the Calculus of Variations introduced in this Chapter present a mathematical procedure for transforming the problem of minimizing an integral to the problem of finding the solution to an ordinary differential equation for $y(x)$. The mathematical foundations of the Calculus of Variations were developed by Leonhard Euler (1707-1783) and Joseph-Louis Lagrange (1736-1813), who developed the mathematical method for finding curves that minimize (or maximize) certain integrals.

1.1.2 Methods of the Calculus of Variations

1.1.2.1 Euler's First Equation

The methods of the Calculus of Variations transform the problem of minimizing (or maximizing) an integral of the form

$$\mathcal{F}[y] = \int_a^b F(y, y'; x) \, dx \qquad (1.2)$$

(with fixed boundary points a and b) into the solution of a differential equation for $y(x)$ expressed in terms of derivatives of the integrand $F(y, y'; x)$, which is assumed to be a smooth function of $y(x)$ and its first derivative $y'(x)$, with a possible explicit dependence on x.

The problem of *extremizing* the integral (1.2) will be treated in analogy with the problem of finding the extremal value of any (smooth) function $f(x)$, i.e., finding the value x_0 such that

$$f'(x_0) = \lim_{\epsilon \to 0} \frac{1}{\epsilon} \left(f(x_0 + \epsilon) - f(x_0) \right) \equiv \frac{1}{h} \left(\frac{d}{d\epsilon} f(x_0 + \epsilon h) \right)_{\epsilon=0} = 0,$$

where $h \neq 0$ is an arbitrary constant factor.[1] First, we introduce the first-order *functional variation* $\delta \mathcal{F}[y; \delta y]$ defined as

$$\delta \mathcal{F}[y; \delta y] \equiv \left(\frac{d}{d\epsilon} \mathcal{F}[y + \epsilon \, \delta y] \right)_{\epsilon=0}$$

$$= \left[\frac{d}{d\epsilon} \left(\int_a^b F\left(y + \epsilon \, \delta y, y' + \epsilon \, \delta y', x \right) dx \right) \right]_{\epsilon=0}, \qquad (1.3)$$

where $\delta y(x)$ is an arbitrary smooth variation of the path $y(x)$ subject to the boundary conditions $\delta y(a) = 0 = \delta y(b)$, i.e., the end points of the path are not affected by the variation (see Fig. 1.1). By performing the ϵ-derivatives in the functional variation (1.3), which involves partial derivatives of $F(y, y', x)$ with respect to y and y', we find

$$\delta \mathcal{F}[y; \delta y] = \int_a^b \left[\delta y(x) \frac{\partial F}{\partial y(x)} + \delta y'(x) \frac{\partial F}{\partial y'(x)} \right] dx,$$

---

[1] An *extremum* point refers to either the minimum or maximum of a one-variable function. A *critical* point, on the other hand, refers to a point where the gradient of a multi-variable function vanishes. Critical points include minima and maxima as well as saddle points (where the function exhibits maxima in some directions and minima in other directions). A function $y(x)$ is said to be a *stationary* solution of the functional (1.2) if the first variation (1.3) vanishes for all variations δy that satisfy the boundary conditions.

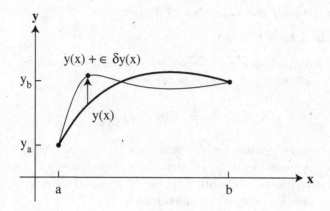

Fig. 1.1   Virtual displacement $\delta y(x)$ for the functional variation (1.3).

which, when the second term is integrated by parts, becomes

$$\delta\mathcal{F}[y;\delta y] = \int_a^b \delta y \left[\frac{\partial F}{\partial y} - \frac{d}{dx}\left(\frac{\partial F}{\partial y'} \right) \right] dx$$
$$+ \left[\delta y_b \left(\frac{\partial F}{\partial y'} \right)_b - \delta y_a \left(\frac{\partial F}{\partial y'} \right)_a \right].$$

Here, since the variation $\delta y(x)$ vanishes at the integration boundaries ($\delta y_b = 0 = \delta y_a$), the last terms involving δy_b and δy_a vanish explicitly and we obtain

$$\delta\mathcal{F}[y;\delta y] = \int_a^b \delta y \left[\frac{\partial F}{\partial y} - \frac{d}{dx}\left(\frac{\partial F}{\partial y'} \right) \right] dx \equiv \int_a^b \delta y \, \frac{\delta\mathcal{F}}{\delta y} \, dx, \quad (1.4)$$

where $\delta\mathcal{F}/\delta y$ is called the *functional derivative* of $\mathcal{F}[y]$ with respect to the function y. The stationarity condition

$$\delta\mathcal{F}[y;\delta y] = 0 \qquad (1.5)$$

for all variations δy yields Euler's First equation

$$\frac{d}{dx}\left(\frac{\partial F}{\partial y'} \right) \equiv y'' \frac{\partial^2 F}{\partial y' \, \partial y'} + y' \frac{\partial^2 F}{\partial y \, \partial y'} + \frac{\partial^2 F}{\partial x \, \partial y'} = \frac{\partial F}{\partial y}, \qquad (1.6)$$

which represents a second-order ordinary differential equation for $y(x)$. According to the Calculus of Variations, the solution $y(x)$ to this ordinary differential equation, subject to the boundary conditions $y(a) = y_a$ and $y(b) = y_b$, yields a solution to the problem of minimizing (or maximizing) the integral (1.2). Lastly, we note that Lagrange's variation operator δ, while analogous to the derivative operator d, commutes with the integral operator, i.e.,

$$\delta \int_a^b P(y(x)) \, dx = \int_a^b P'(y(x)) \, \delta y(x) \, dx,$$

for any smooth function P.

1.1.2.2 *Extremal Values of an Integral*

Euler's First Equation (1.6), which results from the stationarity condition (1.5), does not necessarily imply that the Euler path $y(x)$, in fact, minimizes the integral (1.2). To investigate whether the path $y(x)$ actually minimizes Eq. (1.2), we must evaluate the second-order functional variation

$$\delta^2 \mathcal{F}[y; \delta y] \equiv \left(\frac{d^2}{d\epsilon^2} \mathcal{F}[y + \epsilon \, \delta y] \right)_{\epsilon=0},$$

and investigate its sign. By following steps similar to the derivation of Eq. (1.4), the second-order variation is expressed as

$$\delta^2 \mathcal{F}[y; \delta y] = \int_a^b \left\{ \delta y^2 \left[\frac{\partial^2 F}{\partial y^2} - \frac{d}{dx} \left(\frac{\partial^2 F}{\partial y \partial y'} \right) \right] + (\delta y')^2 \frac{\partial^2 F}{\partial (y')^2} \right\} dx, \tag{1.7}$$

after integration by parts was performed. The necessary and sufficient condition for a minimum is $\delta^2 \mathcal{F} > 0$ and, thus, the sufficient conditions for a minimal integral are

$$\frac{\partial^2 F}{\partial y^2} - \frac{d}{dx} \left(\frac{\partial^2 F}{\partial y \partial y'} \right) > 0 \quad \text{and} \quad \frac{\partial^2 F}{(\partial y')^2} > 0, \tag{1.8}$$

for all smooth variations $\delta y(x)$. For a small enough interval (a, b), the $(\delta y')^2$-term will normally dominate over the $(\delta y)^2$-term and the sufficient condition becomes $\partial^2 F / (\partial y')^2 > 0$ (Legendre's Condition [6]).

Because variational problems often involve finding the minima or maxima of certain integrals, the methods of the Calculus of Variations enable us to find extremal solutions $y_0(x)$ for which the integral $\mathcal{F}[y]$ is stationary (i.e., $\delta \mathcal{F}[y_0] = 0$), without specifying whether the second-order variation is positive-definite (corresponding to a minimum), negative-definite (corresponding to a maximum), or with indefinite sign (i.e., when the coefficients of $(\delta y)^2$ and $(\delta y')^2$ have opposite signs).

1.1.2.3 *Jacobi Equation**

Carl Gustav Jacobi (1804-1851) derived a useful differential equation describing the deviation $u(x) = \overline{y}(x) - y(x)$ between two extremal curves that solve Euler's First Equation (1.6) for a given function $F(x, y, y')$. Upon Taylor expanding Euler's First Equation (1.6) for $\overline{y} = y + u$ and keeping only linear terms in u (which is assumed to be small), we easily obtain the linear ordinary differential equation

$$\frac{d}{dx}\left(u' \frac{\partial^2 F}{(\partial y')^2} + u \frac{\partial^2 F}{\partial y \partial y'}\right) = u \frac{\partial^2 F}{\partial y^2} + u' \frac{\partial^2 F}{\partial y' \partial y}. \tag{1.9}$$

By performing the x-derivative on the second term on the left side, we obtain a partial cancellation with the second term on the right side and obtain the Jacobi equation [6]

$$\frac{d}{dx}\left(\frac{\partial^2 F}{(\partial y')^2} \frac{du}{dx}\right) = u \left[\frac{\partial^2 F}{\partial y^2} - \frac{d}{dx}\left(\frac{\partial^2 F}{\partial y \partial y'}\right)\right]. \tag{1.10}$$

We immediately see that the extremal properties (1.8) of the solutions of Euler's First Equation (1.6) are intimately connected to the behavior of the deviation $u(x)$ between two nearby extremal curves.

We note that the differential equation (1.9) may be derived from the variational principle $\delta \int J(u, u') \, dx = 0$ as the Jacobi-Euler equation

$$\frac{d}{dx}\left(\frac{\partial J}{\partial u'}\right) = \frac{\partial J}{\partial u}, \tag{1.11}$$

where the Jacobi function $J(u, u'; x)$ is defined as

$$J(u, u') \equiv \frac{1}{2}\left(\frac{d^2}{d\epsilon^2} F(y + \epsilon u, y' + \epsilon u')\right)_{\epsilon=0}$$

$$\equiv \frac{u^2}{2} \frac{\partial^2 F}{\partial y^2} + u u' \frac{\partial^2 F}{\partial y \partial y'} + \frac{u'^2}{2} \frac{\partial^2 F}{(\partial y')^2}. \tag{1.12}$$

For example, for $F(y, y') = \sqrt{1 + (y')^2}$, we find $\partial^2 F/\partial y^2 = 0 = \partial^2 F/\partial y \partial y'$ and $\partial^2 F/\partial(y')^2 = \Lambda^{-3}$, where $\Lambda \equiv \sqrt{1 + m^2}$ for the extremal solution $y(x) = m\,x$. The Jacobi function (1.12) for this case is $J(u, u') = \frac{1}{2}(u')^2/\Lambda^3$ and the Jacobi equation (1.10) becomes $(\Lambda^{-3}\,u')' = 0$, or $u'' = 0$ (i.e., deviations diverge linearly).

Lastly, the second functional variation (1.7) can be combined with the Jacobi equation (1.10) to yield the expression [6]

$$\delta^2 \mathcal{F}[y; \delta y] = \int_a^b \frac{\partial^2 F}{(\partial y')^2}\left(\delta y' - \delta y \frac{u'}{u}\right)^2 dx, \tag{1.13}$$

where $u(x)$ is a solution of the Jacobi equation (1.10). We note that the minimum condition $\delta^2 \mathcal{F} > 0$ is now clearly associated with the condition $\partial^2 F/\partial(y')^2 > 0$. Furthermore, we note that the Jacobi equation describing space-time geodesic deviations plays a fundamental role in Einstein's Theory of General Relativity. We shall return to the Jacobi equation in Sec. 1.4, where we briefly discuss Fermat's Principle of Least Time and its applications to the general theory of geometric optics.

1.1.2.4 *Euler's Second Equation*

Under the special condition $\partial F/\partial x \equiv 0$, we may obtain a partial solution to Euler's First Equation (1.6) as follows. First, we write the exact x-derivative of $F(y, y'; x)$ as

$$\frac{dF}{dx} = \frac{\partial F}{\partial x} + y' \frac{\partial F}{\partial y} + y'' \frac{\partial F}{\partial y'},$$

and substitute Euler's First Equation (1.6) to combine the last two terms so that we obtain Euler's Second equation

$$\frac{d}{dx} \left(F - y' \frac{\partial F}{\partial y'} \right) = \frac{\partial F}{\partial x}. \tag{1.14}$$

This equation is especially useful when the integrand $F(y, y')$ in Eq. (1.2) is independent of x, for which Eq. (1.14) yields the solution

$$F(y, y') - y' \frac{\partial F}{\partial y'}(y, y') = \alpha, \tag{1.15}$$

where the constant α is determined from the conditions $y(x_0) = y_0$ and $y'(x_0) = y_0'$. Here, Eq. (1.15) is a *partial* solution (in some sense) of Eq. (1.6), since we have reduced the derivative order from a second-order derivative $y''(x)$ in Eq. (1.6) to a first-order derivative $y'(x)$ in Eq. (1.15) on the solution $y(x)$. Hence, Euler's Second Equation has produced an equation of the form

$$G(y, y'; \alpha) \equiv F(y, y') - y' \frac{\partial F}{\partial y'}(y, y') - \alpha = 0,$$

which can often be integrated by *quadrature* (as we shall see later) by solving for y' as a function of y.

1.1.3 *Path of Shortest Distance and Geodesic Equation*

We now return to the problem of minimizing the length integral (1.1), with the integrand written as $F(y, y') = \sqrt{1 + (y')^2}$. Here, Euler's First

Equation (1.6) yields

$$\frac{d}{dx}\left(\frac{\partial F}{\partial y'}\right) = \frac{y''}{[1+(y')^2]^{3/2}} = \frac{\partial F}{\partial y} = 0,$$

so that the function $y(x)$ that minimizes the length integral (1.1) is the solution of the differential equation $y''(x) = 0$ subject to the boundary conditions $y(0) = 0$ and $y(1) = m$, i.e., the extremal solution is $y(x) = m\,x$. Note that the integrand $F(y, y')$ also satisfies the sufficient minimum conditions (1.8) so that the path $y(x) = m\,x$ is indeed the path of shortest distance between two points on the plane.

1.1.3.1 *Geodesic equation**

We generalize the problem of finding the path of shortest distance on the Euclidean plane (x, y) to the problem of finding *geodesic* paths in arbitrary geometry because it introduces important geometric concepts in Classical Mechanics needed in later chapters. For this purpose, let us consider a path in n-dimensional space from point $\mathbf{x}_A$ to point $\mathbf{x}_B$ parameterized by the continuous parameter σ: $\mathbf{x}(\sigma)$ such that $\mathbf{x}(A) = \mathbf{x}_A$ and $\mathbf{x}(B) = \mathbf{x}_B$. The length integral from point A to B is

$$\mathcal{L}[\mathbf{x}] = \int_A^B \left(g_{ij}\frac{dx^i}{d\sigma}\frac{dx^j}{d\sigma}\right)^{1/2} d\sigma, \tag{1.16}$$

where the space metric g_{ij} is defined so that the squared infinitesimal length element is $ds^2 \equiv g_{ij}(\mathbf{x})\,dx^i\,dx^j$ (summation over repeated indices is implied throughout the text).

Next, using the definition (1.3), the first-order variation $\delta\mathcal{L}[\mathbf{x}]$ is given as

$$\delta\mathcal{L}[\mathbf{x}] = \frac{1}{2}\int_A^B \left[\frac{\partial g_{ij}}{\partial x^k}\delta x^k\frac{dx^i}{d\sigma}\frac{dx^j}{d\sigma} + 2\,g_{ij}\frac{d\delta x^i}{d\sigma}\frac{dx^j}{d\sigma}\right]\frac{d\sigma}{ds/d\sigma}$$

$$= \frac{1}{2}\int_a^b \left[\frac{\partial g_{ij}}{\partial x^k}\delta x^k\frac{dx^i}{ds}\frac{dx^j}{ds} + 2\,g_{ij}\frac{d\delta x^i}{ds}\frac{dx^j}{ds}\right]ds,$$

where $a = s(A)$ and $b = s(B)$ and we have performed a parameterization change: $\mathbf{x}(\sigma) \to \mathbf{x}(s)$. By integrating the second term by parts (with $\delta\mathbf{x}$ vanishing at the end points), we obtain

$$\delta\mathcal{L}[\mathbf{x}] = -\int_a^b \left[\frac{d}{ds}\left(g_{ij}\frac{dx^j}{ds}\right) - \frac{1}{2}\frac{\partial g_{jk}}{\partial x^i}\frac{dx^j}{ds}\frac{dx^k}{ds}\right]\delta x^i\,ds \tag{1.17}$$

$$= -\int_a^b \left[g_{ij}\frac{d^2x^j}{ds^2} + \left(\frac{\partial g_{ij}}{\partial x^k} - \frac{1}{2}\frac{\partial g_{jk}}{\partial x^i}\right)\frac{dx^j}{ds}\frac{dx^k}{ds}\right]\delta x^i\,ds.$$

We now note that, using symmetry properties under interchange of the j-k indices, the second term in Eq. (1.17) can also be written as

$$\left(\frac{\partial g_{ij}}{\partial x^k} - \frac{1}{2}\frac{\partial g_{jk}}{\partial x^i}\right)\frac{dx^j}{ds}\frac{dx^k}{ds} = \frac{1}{2}\left(\frac{\partial g_{ij}}{\partial x^k} + \frac{\partial g_{ik}}{\partial x^j} - \frac{\partial g_{jk}}{\partial x^i}\right)\frac{dx^j}{ds}\frac{dx^k}{ds}$$

$$= \Gamma_{i|jk}\frac{dx^j}{ds}\frac{dx^k}{ds},$$

using the definition of the Christoffel symbol

$$\Gamma^\ell_{jk} = g^{\ell i}\,\Gamma_{i|jk} = \frac{1}{2}g^{\ell i}\left(\frac{\partial g_{ij}}{\partial x^k} + \frac{\partial g_{ik}}{\partial x^j} - \frac{\partial g_{jk}}{\partial x^i}\right) \equiv \Gamma^i_{kj}, \qquad (1.18)$$

where g^{ij} denotes a component of the inverse metric (i.e., $g^{ij}\,g_{jk} = \delta^i{}_k$). Hence, the first-order variation (1.17) can be expressed as

$$\delta\mathcal{L}[\mathbf{x}] = \int_a^b \left[\frac{d^2x^i}{ds^2} + \Gamma^i_{jk}\frac{dx^j}{ds}\frac{dx^k}{ds}\right] g_{i\ell}\,\delta x^\ell\,ds. \qquad (1.19)$$

The stationarity condition $\delta\mathcal{L} = 0$ for arbitrary variations δx^ℓ yields an equation for the path $\mathbf{x}(s)$ of shortest distance known as the *geodesic* equation

$$\frac{d^2x^i}{ds^2} + \Gamma^i_{jk}\frac{dx^j}{ds}\frac{dx^k}{ds} = 0. \qquad (1.20)$$

Returning to two-dimensional Euclidean geometry, where the components of the metric tensor are constants (i.e., $ds^2 = dx^2 + dy^2$), the geodesic equations are $x''(s) = 0 = y''(s)$, which once again leads to a straight line.

1.1.3.2 *Geodesic equation on a sphere*

For example, geodesic curves on the surface of a sphere of radius R are expressed in terms of extremal curves of the length functional

$$\mathcal{L}[\varphi] = \int R\sqrt{1 + \sin^2\theta\left(\frac{d\varphi}{d\theta}\right)^2}\,d\theta \equiv R\int L(\varphi',\theta)\,d\theta, \qquad (1.21)$$

where the azimuthal angle $\varphi(\theta)$ is an arbitrary function of the polar angle θ. Since the function $L(\varphi',\theta)$ in Eq. (1.21) is independent of the azimuthal angle φ, its corresponding Euler's First Equation is

$$\frac{\partial L}{\partial\varphi'} = \frac{\sin^2\theta\,\varphi'}{\sqrt{1 + \sin^2\theta\,(\varphi')^2}} = \sin\alpha,$$

where α is an arbitrary constant angle. Solving for φ' we find

$$\varphi'(\theta) = \frac{\sin\alpha}{\sin\theta\,\sqrt{\sin^2\theta - \sin^2\alpha}},$$

which can, thus, be integrated to give

$$\varphi - \beta = \int \frac{\sin\alpha \, d\theta}{\sin\theta \sqrt{\sin^2\theta - \sin^2\alpha}} = -\int \frac{\tan\alpha \, du}{\sqrt{1 - u^2 \tan^2\alpha}},$$

where β is another constant angle and we used the change of variable $u = \cot\theta$. A simple trigonometric substitution finally yields

$$\cos(\varphi - \beta) = \tan\alpha \cot\theta, \tag{1.22}$$

which describes a great circle on the surface of the sphere. We easily verify this statement by converting Eq. (1.22) into the equation for a plane that passes through the origin:

$$z \sin\alpha = x \cos\alpha \cos\beta + y \cos\alpha \sin\beta.$$

The intersection of this plane with the unit sphere is expressed in terms of the coordinate functions $x = \sin\alpha \cos\beta$, $y = \sin\alpha \sin\beta$, and $z = \cos\alpha$.

1.2 Classical Variational Problems

The development of the Calculus of Variations led to the resolution of several classical optimization problems in mathematics and physics. In this Section, we present two classical variational problems that are connected to its original development. First, in the isoperimetric problem, we show how Lagrange modified Euler's formulation of the Calculus of Variations by allowing constraints to be imposed on the search for finding extremal values of certain integrals. Next, in the brachistochrone problem, we show how the Calculus of Variations is used to find the path of *quickest* descent for a bead sliding along a frictionless wire under the action of gravity.

1.2.1 *Isoperimetric Problem*

Isoperimetric problems represent some of the earliest applications of the variational approach to solving mathematical optimization problems. Pappus (ca. 290-350) was among the first to recognize that among all the isoperimetric closed planar curves (i.e., closed curves that have the same perimeter length), the circle encloses the greatest area.[2] The variational

---

[2]Such results are normally described in terms of the so-called isoperimetric inequalities $4\pi A \leq L^2$, where A denotes the area enclosed by a closed curve of perimeter length L; here, equality is satisfied by the circle.

formulation of the (planar) isoperimetric problem requires that we maximize the area integral $A = \int y(x)\, dx$ while keeping the perimeter length integral $L = \int \sqrt{1 + (y')^2}\, dx$ constant.

The isoperimetric problem falls in a class of variational problems called *constrained* variational principles, where a certain functional $\int f(y, y', x)\, dx$ is to be optimized under the constraint that another functional $\int g(y, y', x)\, dx$ be held constant (say at value G). The constrained variational principle is then expressed in terms of the functional

$$\mathcal{F}_\lambda[y] = \int f(y, y', x)\, dx + \lambda \left(G - \int g(y, y', x)\, dx \right)$$

$$= \int \left[f(y, y', x) - \lambda\, g(y, y', x) \right] dx + \lambda G, \qquad (1.23)$$

where the parameter λ is called a Lagrange *multiplier*. Note that the functional $\mathcal{F}_\lambda[y]$ is chosen, on the one hand, so that the derivative

$$\frac{d\mathcal{F}_\lambda[y]}{d\lambda} = G - \int g(y, y', x)\, dx = 0$$

enforces the constraint for all curves $y(x)$. On the other hand, the stationarity condition $\delta \mathcal{F}_\lambda = 0$ for the functional (1.23) with respect to arbitrary variations $\delta y(x)$ (which vanish at the integration boundaries) yields Euler's First Equation:

$$\frac{d}{dx}\left(\frac{\partial f}{\partial y'} - \lambda \frac{\partial g}{\partial y'} \right) = \frac{\partial f}{\partial y} - \lambda \frac{\partial g}{\partial y}. \qquad (1.24)$$

Here, we assume that this second-order differential equation is to be solved subject to the conditions $y(x_0) = y_0$ and $y'(x_0) = 0$; the solution $y(x; \lambda)$ of Eq. (1.24) is, however, parameterized by the Lagrange multiplier λ.

If the integrands $f(y, y')$ and $g(y, y')$ in Eq. (1.23) are both explicitly independent of x, then Euler's Second Equation (1.15) for the functional (1.23) becomes

$$\frac{d}{dx}\left[\left(f - y' \frac{\partial f}{\partial y'} \right) - \lambda \left(g - y' \frac{\partial g}{\partial y'} \right) \right] = 0. \qquad (1.25)$$

By integrating this equation we obtain

$$\left(f - y' \frac{\partial f}{\partial y'} \right) - \lambda \left(g - y' \frac{\partial g}{\partial y'} \right) = 0,$$

where the constant of integration on the right is chosen from the conditions $y(x_0) = y_0$ and $y'(x_0) \equiv 0$ (i.e., x_0 is an extremum point of $y(x)$), so that the value of the constant Lagrange multiplier is now defined as $\lambda =$

$f(y_0, 0)/g(y_0, 0)$. Hence, the solution $y(x; \lambda)$ of the constrained variational problem (1.23) is now uniquely determined.

We return to the isoperimetric problem now represented in terms of the constrained functional

$$
\begin{aligned}
\mathcal{A}_\lambda[y] &= \int y \, dx + \lambda \left(L - \int \sqrt{1 + (y')^2} \, dx \right) \\
&= \int \left[y - \lambda \sqrt{1 + (y')^2} \right] dx + \lambda L,
\end{aligned}
\tag{1.26}
$$

where L denotes the value of the constant-length constraint. From Eq. (1.24), the stationarity of the functional (1.26) with respect to arbitrary variations $\delta y(x)$ yields

$$
\frac{d}{dx} \left(-\frac{\lambda y'}{\sqrt{1 + (y')^2}} \right) = 1,
$$

which can be integrated to give

$$
-\frac{\lambda y'}{\sqrt{1 + (y')^2}} = x - x_0,
\tag{1.27}
$$

where x_0 denotes a constant of integration associated with $y'(x_0) = 0$. Since the integrands $f(y, y') = y$ and $g(y, y') = \sqrt{1 + (y')^2}$ are both explicitly independent of x, then Euler's Second Equation (1.25) applies, and we obtain

$$
\frac{d}{dx} \left(y - \frac{\lambda}{\sqrt{1 + (y')^2}} \right) = 0,
$$

which can be integrated to give

$$
\frac{\lambda}{\sqrt{1 + (y')^2}} = y.
\tag{1.28}
$$

Lastly, the constant of integration is chosen with $y'(x_0) = 0$ so that $y(x_0) = \lambda$. By combining Eqs. (1.27) and (1.28), we obtain $y y' + (x - x_0) = 0$, which can be integrated to give $y^2(x) = \lambda^2 - (x - x_0)^2$. We immediately recognize that the maximal isoperimetric curve $y(x)$ is a circle of radius $r = \lambda$ with perimeter length $L = 2\pi \lambda$ and maximal enclosed area $A = \pi \lambda^2 = L^2/4\pi$.

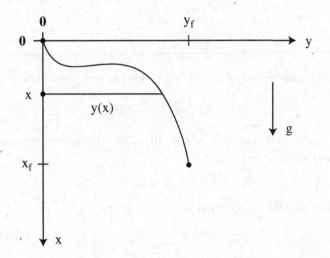

Fig. 1.2 Brachistochrone problem.

1.2.2 *Brachistochrone Problem*

The brachistochrone problem is a *least-time* variational problem, which was first solved in 1696 by Jean (Johann) Bernoulli (1667-1748). The problem can be stated as follows. A bead is released from rest (at the origin in Fig. 1.2) and slides down a frictionless wire that connects the origin to a given point (x_f, y_f). The question posed by the brachistochrone problem is to determine the shape $y(x)$ of the wire for which the frictionless descent of the bead under gravity takes the shortest amount of time.

Using the (x, y)-coordinates set up in Fig. 1.2, the speed of the bead after it has fallen a vertical distance x along the wire is $v = \sqrt{2g\,x}$ (where g denotes the gravitational acceleration) and, thus, the time integral

$$T[y] = \int \frac{ds}{v} = \int_0^{x_f} \sqrt{\frac{1 + (y')^2}{2\,gx}}\, dx = \int_0^{x_f} F(y, y', x)\, dx, \quad (1.29)$$

is a functional of the path $y(x)$. Note that, in the absence of friction, the bead's mass does not enter into the problem. Since the integrand of Eq. (1.29) is independent of the y-coordinate ($\partial F/\partial y = 0$), Euler's First Equation (1.6) simply yields

$$\frac{d}{dx}\left(\frac{\partial F}{\partial y'}\right) = 0 \quad \rightarrow \quad \frac{\partial F}{\partial y'} = \frac{y'}{\sqrt{2\,gx\,[1 + (y')^2]}} = \alpha,$$

where α is a constant, which can be rewritten in terms of the scale length

$\ell = (2\alpha^2 g)^{-1}$ as

$$\frac{(y')^2}{1 + (y')^2} = \frac{x}{\ell}. \tag{1.30}$$

Integration by quadrature of Eq. (1.30) yields the integral solution

$$y(x) = \int_0^x \sqrt{\frac{s}{\ell - s}}\, ds,$$

subject to the initial condition $y(x = 0) = 0$. Using the trigonometric substitution (with $\ell = 2a$)

$$s = 2a\, \sin^2(\theta/2) = a\,(1 - \cos\theta),$$

we obtain the parametric solution

$$x(\theta) = a\,(1 - \cos\theta) \tag{1.31}$$

and

$$y(\theta) = \int_0^\theta \sqrt{\frac{1 - \cos\theta}{1 + \cos\theta}}\, a\, \sin\theta\, d\theta$$

$$= a \int_0^\theta (1 - \cos\theta)\, d\theta = a\,(\theta - \sin\theta). \tag{1.32}$$

This solution yields a parametric representation of the *cycloid* (Fig. 1.3) where the bead is placed on a rolling hoop of radius a.

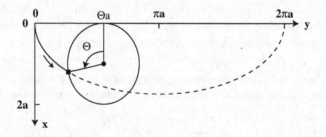

Fig. 1.3　Brachistochrone solution.

1.3　Fermat's Principle of Least Time

Several *minimum* principles have been invoked throughout the history of Physics to explain the behavior of light and particles. In one of its earliest form, Hero of Alexandria (ca. 75 AD) stated that *light travels in a straight line* and that light follows a path of shortest distance when it is reflected. In 1657, Pierre de Fermat (1601-1665) stated the Principle of *Least Time*, whereby light travels between two points along a path that minimizes the travel time, to explain Snell's Law (Willebrord Snell, 1591-1626) associated with light refraction in a stratified medium. Using the index of refraction $n_0 \geq 1$ of the uniform medium, the speed of light in the medium is expressed as $v_0 = c/n_0 \leq c$, where c is the speed of light in vacuum. This straight path in a uniform medium is not only a path of shortest distance but also a path of least time.

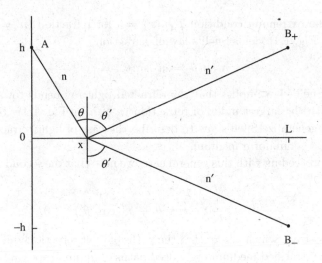

Fig. 1.4   Reflection path (AxB_+) and refraction path (AxB_-) for light propagating in a stratified medium.

The laws of reflection and refraction as light propagates in uniform media separated by sharp boundaries (see segments AxB_+ and AxB_- in Fig. 1.4) are easily formulated as minimization problems as follows. The time taken by light to go from point $A = (0, h)$ to point $B_\pm = (L, \pm h)$ after being reflected or refracted at point $(x, 0)$ is given by

$$T_{AB}(x) = c^{-1} \left[n \sqrt{x^2 + h^2} + n' \sqrt{(L-x)^2 + h^2} \right],$$

where n and n' denote the indices of refraction of the medium along path Ax and $xB_\pm$, respectively. We easily evaluate the derivative of $T_{AB}(x)$ to find

$$\frac{dT_{AB}(x)}{dx} = c^{-1}\left[\, n\,\frac{x}{\sqrt{x^2 + h^2}} \;-\; n'\,\frac{(L-x)}{\sqrt{(L-x)^2 + h^2}} \,\right]$$

$$\equiv c^{-1}\left(n\,\sin\theta \;-\; n'\,\sin\theta'\right), \tag{1.33}$$

where the angles θ and θ' are defined in Fig. 1.4. Here, the law of reflection ($n' = n$ and $B = B_+$ in Fig. 1.4) is expressed in terms of the extremum condition $T'_{AB}(x) = 0$, which implies that the path of least time is obtained when the reflected angle θ' is equal to the incidence angle θ (or $x = L/2$ in Fig. 1.4).

Next, the extremum condition $T'_{AB}(x) = 0$ for refraction ($n' \neq n$ and $B = B_-$ in Fig. 1.4) yields Snell's law of refraction

$$n\,\sin\theta \;=\; n'\,\sin\theta'. \tag{1.34}$$

Note that Snell's law implies that the refracted light ray bends toward the medium with the largest index of refraction ($n > n'$ in Fig. 1.4). In what follows, we generalize Snell's law to describe the case of light refraction in a continuous nonuniform medium.

Before proceeding with this general case, we note that the second derivative

$$\frac{d^2T_{AB}(x)}{dx^2} \;=\; \frac{1}{hc}\left(n\,\cos^3\theta \;+\; n'\,\cos^3\theta'\right) \;>\; 0,$$

is strictly positive, which proves that the paths of light reflection and refraction in a flat stratified medium are indeed paths of minimal optical lengths. Note that for some curved reflecting surfaces, however, the reflected path corresponds to a path of maximum optical length (see problem 16). This example emphasizes the fact that Fermat's Principle is in fact a principle of *stationary* time.

1.3.1   *Light Propagation in a Nonuniform Medium*

According to Fermat's Principle, light propagates in a nonuniform medium by traveling along a path that *minimizes* the travel time between an initial point A (where a light ray is launched) and a final point B (where the light ray is received). The time taken by a light ray following a path γ from point A to point B (parameterized by σ) is [3]

$$T[\mathbf{x}] \;=\; \int c^{-1}n(\mathbf{x})\,\left|\frac{d\mathbf{x}}{d\sigma}\right|\,d\sigma \;=\; c^{-1}\,\mathcal{L}_n[\mathbf{x}], \tag{1.35}$$

where $\mathcal{L}_n[\mathbf{x}]$ represents the length of the *optical* path taken by light as it travels in a nonuniform medium with refractive index $n(\mathbf{x})$ and

$$\left|\frac{d\mathbf{x}}{d\sigma}\right| = \sqrt{\left(\frac{dx}{d\sigma}\right)^2 + \left(\frac{dy}{d\sigma}\right)^2 + \left(\frac{dz}{d\sigma}\right)^2}. \tag{1.36}$$

Fermat's Principle of Least Time states that light traveling in a nonuniform medium follows an optical path $\mathbf{x}(\sigma)$ that is a stationary solution of the variational principle

$$\delta \mathcal{T}[\mathbf{x}] \equiv 0. \tag{1.37}$$

We now consider ray propagation in two dimensions (x, y), with the index of refraction $n(y)$, and return to general properties of ray propagation in Sec. 1.4.

For ray propagation in two dimensions (labeled x and y) in a medium with nonuniform refractive index $n(y)$, an arbitrary point $(x,\ y = y(x))$ along the light path $\mathbf{x}(\sigma)$ is parameterized by the x-coordinate [i.e., $\sigma = x$ in Eq. (1.35)], which starts at point $A = (a, y_a)$ and ends at point $B = (b, y_b)$. Along the path $y : x \mapsto y(x)$, the infinitesimal length element is $ds = \sqrt{1 + (y')^2}\, dx$ and the optical length

$$\mathcal{L}_n[y] = \int_a^b n(y)\ \sqrt{1 + (y')^2}\ dx \tag{1.38}$$

is a *functional* of y (i.e., changing the path y changes the value of the integral $\mathcal{L}_n[y]$).

We now apply the variational principle (1.37) for the case where $F(y, y') = n(y)\ \sqrt{1 + (y')^2}$, from which we find

$$\frac{\partial F}{\partial y'} = \frac{n(y)\, y'}{\sqrt{1 + (y')^2}} \quad \text{and} \quad \frac{\partial F}{\partial y} = n'(y)\ \sqrt{1 + (y')^2},$$

so that Euler's First Equation (1.6) becomes

$$n(y)\, y'' = n'(y)\ \left[1 + (y')^2\right]. \tag{1.39}$$

Although the solution of this (nonlinear) second-order ordinary differential equation is difficult to obtain for general functions $n(y)$, we can nonetheless obtain a qualitative picture of its solution by noting that y'' has the same sign as $n'(y)$. Hence, when $n'(y) = 0$ for all y (i.e., the medium is spatially uniform), the solution $y'' = 0$ yields the straight line $y(x; \varphi_0) = \tan \varphi_0\ x$, where φ_0 denotes the initial launch angle (as measured from the horizontal axis). The case where $n'(y) > 0$ (or < 0), on the other hand, yields a light path which is concave upward (or downward) as will be shown below.

Note that the sufficient conditions (1.8) for a minimal optical path are expressed as

$$\frac{\partial^2 F}{(\partial y')^2} = \frac{n}{[1 + (y')^2]^{3/2}} > 0,$$

which is satisfied for all refractive media, and

$$\frac{\partial^2 F}{\partial y^2} - \frac{d}{dx}\left(\frac{\partial^2 F}{\partial y \partial y'}\right) = n'' \sqrt{1 + (y')^2} - \frac{d}{dx}\left(\frac{n' y'}{\sqrt{1 + (y')^2}}\right)$$

$$= \frac{n^2}{F}\frac{d^2 \ln n}{dy^2},$$

whose sign is indefinite. Hence, the sufficient condition for a minimal optical length for light traveling in a nonuniform refractive medium is $d^2 \ln n / dy^2 > 0$; note, however, that only the stationarity of the optical path is physically meaningful and, thus, we shall not discuss the minimal properties of light paths in what follows.

Since the function $F(y, y') = n(y) \sqrt{1 + (y')^2}$ is explicitly independent of x, Euler's Second Equation yields

$$F - y'\frac{\partial F}{\partial y'} = \frac{n(y)}{\sqrt{1 + (y')^2}} = \text{constant},$$

and, thus, the partial solution of Eq. (1.39) is

$$n(y) = N \sqrt{1 + (y')^2}, \tag{1.40}$$

where N is a constant determined from the initial conditions of the light ray. We note that Eq. (1.40) states that as a light ray enters a region of increased (decreased) refractive index, the slope of its path also increases (decreases). In particular, by substituting Eq. (1.39) into Eq. (1.40), we find $N^2 y'' = n(y) n'(y)$, and, hence, the path of a light ray is concave upward (downward) where $n'(y)$ is positive (negative), as previously discussed. Eq. (1.40) can be integrated by *quadrature* to give the integral solution

$$x(y) = \int_0^y \frac{N \, ds}{\sqrt{[n(s)]^2 - N^2}}, \tag{1.41}$$

subject to the condition $x(y = 0) = 0$. From the explicit dependence of the index of refraction $n(y)$, we may be able to perform the integration in Eq. (1.41) to obtain $x(y)$ and, thus, obtain an explicit solution $y(x)$ by inverting $x(y)$.

1.3.2 Snell's Law

We now show that the partial solution (1.40) corresponds to Snell's Law for light refraction in a nonuniform medium. Consider a light ray traveling in the (x, y)-plane launched from the initial position $(0, 0)$ at an initial angle φ_0 (measured from the x-axis) so that $y'(0) = \tan \varphi_0$ is the slope at $x = 0$. The constant N is then simply determined from Eq. (1.40) as $N = n_0 \cos \varphi_0$, where $n_0 = n(0)$ is the refractive index at $y(0) = 0$. Next, let $y'(x) = \tan \varphi(x)$ be the slope of the light ray at $(x, y(x))$, so that $\sqrt{1 + (y')^2} = \sec \varphi$ and Eq. (1.40) becomes $n(y) \cos \varphi = n_0 \cos \varphi_0$. Lastly, when we substitute the complementary angle $\theta = \pi/2 - \varphi$ (measured from the vertical y-axis), we obtain the *local* form of Snell's Law of refraction

$$n[y(x)] \sin \theta(x) = n_0 \sin \theta_0, \qquad (1.42)$$

properly generalized to include a light path in a nonuniform refractive medium. Note that Snell's Law (1.42) does not tell us anything about the actual light path $y(x)$; this solution must come from solving Eq. (1.41).

1.3.3 Application of Fermat's Principle

As an application of Fermat's Principle in two dimensions, we consider the propagation of a light ray in a medium with linear refractive index $n(y) = n_0 (1 - \beta y)$ exhibiting a constant gradient $n'(y) = -n_0 \beta$. Substituting this profile into the optical-path solution (1.41), we find

$$x(y) = \int_0^y \frac{\cos \varphi_0 \, ds}{\sqrt{(1 - \beta s)^2 - \cos^2 \varphi_0}}. \qquad (1.43)$$

Next, we use the trigonometric substitution

$$y(\varphi) = \frac{1}{\beta} \left(1 - \frac{\cos \varphi_0}{\cos \varphi} \right), \qquad (1.44)$$

with $\varphi = \varphi_0$ at $(x, y) = (0, 0)$, so that Eq. (1.43) becomes

$$x(\varphi) = -\frac{\cos \varphi_0}{\beta} \ln \left(\frac{\sec \varphi + \tan \varphi}{\sec \varphi_0 + \tan \varphi_0} \right). \qquad (1.45)$$

The *parametric* solution (1.44)-(1.45) for the optical path in a linear medium shows that the path reaches a maximum height $\bar{y} = y(0)$ at a distance $\bar{x} = x(0)$ when the *tangent* angle φ is zero:

$$\bar{x} = \frac{\cos \varphi_0}{\beta} \ln(\sec \varphi_0 + \tan \varphi_0) \quad \text{and} \quad \bar{y} = \frac{1 - \cos \varphi_0}{\beta}.$$

Lastly, we note that by expressing $y(x; \beta)$ as a function of x, we obtain

$$y(x; \beta) = \frac{1}{\beta}\left[1 - \cos\varphi_0 \cosh\left(\beta x \sec\varphi_0 - \psi_0\right)\right], \qquad (1.46)$$

where $\psi_0 = \operatorname{arccosh}(\sec\varphi_0) \equiv \beta\bar{x}\sec\varphi_0$. In the uniform limit $(\beta = 0)$, we find the straight-line equation $y(x; 0) \equiv x\tan\varphi_0$.

1.4   Geometric Formulation of Ray Optics*

1.4.1   *Frenet-Serret Curvature of Light Path*

We now return to the general formulation for light-ray propagation based on the time integral (1.35), where the integrand is

$$F\left(\mathbf{x}, \frac{d\mathbf{x}}{d\sigma}\right) = n(\mathbf{x})\left|\frac{d\mathbf{x}}{d\sigma}\right|,$$

and light rays are allowed to travel in a three-dimensional refractive medium with a general index of refraction $n(\mathbf{x})$. Euler's First equation in this case is

$$\frac{d}{d\sigma}\left(\frac{\partial F}{\partial(d\mathbf{x}/d\sigma)}\right) = \frac{\partial F}{\partial\mathbf{x}}, \qquad (1.47)$$

where

$$\frac{\partial F}{\partial(d\mathbf{x}/d\sigma)} = \frac{n}{\Lambda}\frac{d\mathbf{x}}{d\sigma} \quad\text{and}\quad \frac{\partial F}{\partial\mathbf{x}} = \Lambda\,\nabla n,$$

with $\Lambda = |d\mathbf{x}/d\sigma|$ given by Eq. (1.36). Euler's First Equation (1.47), therefore, yields the Euler-Fermat equation

$$\frac{d}{d\sigma}\left(\frac{n}{\Lambda}\frac{d\mathbf{x}}{d\sigma}\right) = \Lambda\,\nabla n. \qquad (1.48)$$

Euler's Second Equation, on the other hand, states that

$$H(\sigma) \equiv F\left(\mathbf{x}, \frac{d\mathbf{x}}{d\sigma}\right) - \frac{d\mathbf{x}}{d\sigma}\cdot\frac{\partial F}{\partial(d\mathbf{x}/d\sigma)} = 0$$

is a constant of motion. Note that, while Euler's Second Equation (1.40) proved very useful in providing an explicit solution (Snell's Law) to finding the optical path in a nonuniform medium with index of refraction $n(y)$, it appears that Euler's Second Equation $H(\sigma) \equiv 0$ now reveals no information about the optical path. Where did the information go? To answer this question, we apply the Euler-Fermat equation (1.48) to the two-dimensional case where $\sigma = x$ and $\Lambda = \sqrt{1 + (y')^2}$ with $\nabla n = n'(y)\,\widehat{y}$. Hence, the Euler-Fermat equation (1.48) becomes

$$\frac{d}{dx}\left[\frac{n}{\Lambda}\,(\widehat{x} + y'\,\widehat{y})\right] = \Lambda\, n'\,\widehat{y},$$

from which we immediately conclude that Euler's Second Equation (1.40), $n = N\Lambda$, now appears as a constant of the motion $d(n/\Lambda)/dx = 0$ associated with a symmetry of the optical medium (i.e., the optical properties of the medium are invariant under translation along the x-axis). The association of symmetries with constants of the motion will later be discussed in terms of Noether's Theorem (see problem 17 and Sec. 2.5).

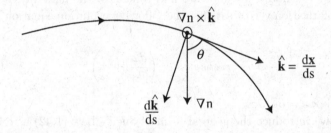

Fig. 1.5   Light-path curvature and the Frenet-Serret frame following a light plane that lies on the surface of the page.

We now look at how the Euler-Fermat equation (1.48) can be simplified by an appropriate choice of parameterization. First, we can choose a ray parameterization $\Lambda = ds/d\sigma \equiv n$, so that the Euler-Fermat equation (1.48) becomes $d^2\mathbf{x}/d\sigma^2 = n\nabla n = \frac{1}{2}\nabla n^2$ and, thus, the light ray is *accelerated* toward regions of higher index of refraction (see Fig. 1.5). Next, by choosing the ray parametrization $d\sigma = ds$ (so that $\Lambda = 1$), the Euler-Fermat equation (1.48) becomes

$$\frac{d}{ds}\left(n\,\frac{d\mathbf{x}}{ds}\right) = \nabla n. \qquad (1.49)$$

Since the ray *velocity* $d\mathbf{x}/ds = \widehat{k}$ is a unit vector, which defines the direction of the wave vector $\mathbf{k}$, Eq. (1.49) yields the *light-curvature* equation

$$\frac{d\widehat{k}}{ds} = \widehat{k} \times \left(\nabla \ln n \times \widehat{k}\right) \equiv \kappa\,\widehat{n}, \qquad (1.50)$$

where $\widehat{n}$ defines the principal normal unit vector and the *Frenet-Serret* curvature κ of the light path is $\kappa = |\nabla \ln n \times \widehat{k}|$ (see Appendix A). Note that for the one-dimensional problem discussed in Sec. 1.3.1, the curvature is $\kappa = |n'|/(n\,\Lambda) = |y''|/\Lambda^3$ in agreement with the Frenet-Serret curvature.

A light wave is characterized by a polarization (unit) vector $\widehat{e}$ that is perpendicular to $\widehat{k}$. We may, thus, write the polarization vector as

$$\widehat{e} \equiv \cos\varphi\,\widehat{n} + \sin\varphi\,\widehat{b}, \tag{1.51}$$

where the normal and binormal unit vectors $\widehat{n}$ and $\widehat{b}$ are perpendicular to the wave-vector $\mathbf{k}$ of a light ray. Using the Frenet-Serret equations

$$\frac{d\widehat{n}}{ds} = \tau\,\widehat{b} - \kappa\,\widehat{k} \text{ and } \frac{d\widehat{b}}{ds} = -\tau\,\widehat{n}, \tag{1.52}$$

where κ and τ denote the curvature and torsion of the light ray, we find that the polarization vector satisfies the following evolution equation along a light ray:

$$\frac{d\widehat{e}}{ds} = -\kappa\,\cos\varphi\,\widehat{k} + \left(\frac{d\varphi}{ds} + \tau\right)\widehat{h}, \tag{1.53}$$

where $\widehat{h} \equiv \widehat{k} \times \widehat{e} = \partial\widehat{e}/\partial\varphi$.

Lastly, we introduce the general form of Snell's Law (1.42) as follows. First, we define the unit vector $\widehat{g} = \nabla n/(|\nabla n|)$ and, after performing the cross-product of Eq. (1.48) with $\widehat{g}$, we obtain the identity

$$\widehat{g} \times \frac{d}{ds}\left(n\,\frac{d\mathbf{x}}{ds}\right) = \widehat{g} \times \nabla n = 0.$$

Using this identity, we readily evaluate the s-derivative of $n\,\widehat{g} \times \widehat{k}$:

$$\frac{d}{ds}\left(\widehat{g} \times n\,\frac{d\mathbf{x}}{ds}\right) = \frac{d\widehat{g}}{ds} \times \left(n\,\frac{d\mathbf{x}}{ds}\right) = \frac{d\widehat{g}}{ds} \times n\,\widehat{k}.$$

Hence, if the unit vector $\widehat{g}$ is constant along the path of a light ray (i.e., $d\widehat{g}/ds = 0$), we then find the conservation law

$$\frac{d}{ds}\left(\widehat{g} \times n\,\widehat{k}\right) = 0, \tag{1.54}$$

which implies that the vector quantity $n\,\widehat{g} \times \widehat{k}$ is a constant along the light path. Note that, when a light ray propagates in two dimensions, this conservation law implies that the quantity $|\widehat{g} \times n\,\widehat{k}| = n\sin\theta$ is also a constant along the light path, where θ is the angle defined as $\cos\theta \equiv \widehat{g} \cdot \widehat{k}$. The conservation law (1.54), therefore, represents a generalization of Snell's Law (1.42).

1.4.2 *Light Propagation in Spherical Geometry*

By using the general ray-orbit equation (1.50), we can also show that, for a spherically-symmetric nonuniform medium with index of refraction $n(r)$, the light-ray orbit $\mathbf{r}(s)$ satisfies the conservation law

$$\frac{d}{ds}\left(\mathbf{r} \times n(r)\,\frac{d\mathbf{r}}{ds}\right) = \mathbf{r} \times \frac{d}{ds}\left(n(r)\,\frac{d\mathbf{r}}{ds}\right) = \mathbf{r} \times \nabla n(r) = 0. \qquad (1.55)$$

Next, we use the fact that the ray-orbit path is planar and, thus, we write

$$\mathbf{r} \times \frac{d\mathbf{r}}{ds} = r\,\sin\varphi\,\widehat{\mathbf{z}}, \qquad (1.56)$$

where φ denotes the angle between the position vector $\mathbf{r}$ and the tangent vector $d\mathbf{r}/ds$ (see Fig. 1.6).

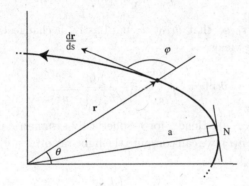

Fig. 1.6 Light path in a nonuniform medium with spherical symmetry.

Using Eq. (1.56), the conservation law (1.55) for ray orbits in a spherically-symmetric medium can, therefore, be expressed as

$$n(r)\,r\,\sin\varphi(r) = N\,a, \qquad (1.57)$$

known as Bouguer's formula (Pierre Bouguer, 1698-1758), where N and a are constants (see Fig. 1.6); note that the condition $n(r)\,r \geq N\,a$ must also be satisfied since $\sin\varphi(r) \leq 1$. This conservation is analogous to the conservation of angular momentum for particles moving in a central-force potential (see Chap. 4).

An explicit expression for the ray orbit $r(\theta)$ is obtained as follows. First, since $d\mathbf{r}/ds$ is a unit vector, we find

$$\frac{d\mathbf{r}}{ds} = \frac{d\theta}{ds}\left(r\,\widehat{\theta} + \frac{dr}{d\theta}\,\widehat{r}\right) = \frac{r\,\widehat{\theta} + (dr/d\theta)\,\widehat{r}}{\sqrt{r^2 + (dr/d\theta)^2}},$$

so that

$$\frac{d\theta}{ds} = \frac{1}{\sqrt{r^2 + (dr/d\theta)^2}}.$$

and Eq. (1.56) yields

$$\mathbf{r} \times \frac{d\mathbf{r}}{ds} = r \sin\varphi\,\widehat{\mathbf{z}} = r^2 \frac{d\theta}{ds}\,\widehat{\mathbf{z}} \;\to\; \sin\varphi = \frac{r}{\sqrt{r^2 + (dr/d\theta)^2}} = \frac{Na}{nr},$$

where we made use of Bouguer's formula (1.57). Next, integration by quadrature yields

$$\theta(r) = N a \int_{r_0}^{r} \frac{d\rho}{\rho\,\sqrt{n^2(\rho)\,\rho^2 - N^2 a^2}},$$

where we choose r_0 so that $\theta(r_0) = 0$. Lastly, a change of integration variable $\eta = Na/\rho$ yields

$$\theta(r) = \int_{Na/r}^{Na/r_0} \frac{d\eta}{\sqrt{\overline{n}^2(\eta) - \eta^2}}, \tag{1.58}$$

where $\overline{n}(\eta) \equiv n(Na/\eta)$. Hence, for a spherically-symmetric medium with index of refraction $n(r)$, we can compute the light-ray orbit $r(\theta)$ by inverting the integral (1.58) for $\theta(r)$.

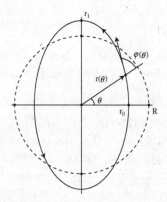

Fig. 1.7 Elliptical light path in a spherically-symmetric refractive medium.

Consider, for example, the spherically-symmetric refractive index $n(r) = n_0\,\sqrt{2 - (r/R)^2}$, where $n_0 = n(R)$ denotes the refractive index at $r = R$. Introducing the dimensional parameter $\epsilon = a/R$ and the transformation $\sigma = \eta^2$, Eq. (1.58) becomes

$$\theta(r) = \int_{Na/r}^{Na/r_0} \frac{\eta \, d\eta}{\sqrt{n_0^2 \left(2\,\eta^2 - N^2 \mathrm{e}^2\right) - \eta^4}}$$

$$= \frac{1}{2} \int_{(Na/r)^2}^{(Na/r_0)^2} \frac{d\sigma}{\sqrt{n_0^4 \, \mathrm{e}^2 - (\sigma - n_0^2)^2}},$$

where $\mathrm{e} = \sqrt{1 - N^2 \epsilon^2 / n_0^2}$ (assuming that $n_0 > N\epsilon$). Next, using the trigonometric substitution $\sigma = n_0^2 \left(1 + \mathrm{e} \cos \chi\right)$, we find $\theta(r) = \frac{1}{2} \chi(r)$ or

$$r(\theta) = \frac{r_0 \sqrt{1 + \mathrm{e}}}{\sqrt{1 + \mathrm{e} \cos 2\theta}},$$

which represents an ellipse (see Fig. 1.7)

$$\left(\frac{x}{R \sqrt{1 - \mathrm{e}}}\right)^2 + \left(\frac{y}{R \sqrt{1 + \mathrm{e}}}\right)^2 = 1$$

with semi-major and semi-minor axes $r_1 = R \left(1 + \mathrm{e}\right)^{1/2}$ and $r_0 = R \left(1 - \mathrm{e}\right)^{1/2}$, respectively. This example shows that, surprisingly, it is possible to trap light!

1.4.3 *Geodesic Representation of Light Propagation*

We now investigate the geodesic properties of light propagation in a nonuniform refractive medium. For this purpose, let us consider a path AB in space from point A to point B parameterized by the continuous parameter σ, i.e., $\mathbf{x}(\sigma)$ such that $\mathbf{x}(A) = \mathbf{x}_A$ and $\mathbf{x}(B) = \mathbf{x}_B$. The time taken by light in propagating from A to B is

$$\mathcal{T}[\mathbf{x}] = \int_A^B \frac{dt}{d\sigma} \, d\sigma = \int_A^B \frac{n}{c} \left(g_{ij} \frac{dx^i}{d\sigma} \frac{dx^j}{d\sigma}\right)^{1/2} \cdot d\sigma, \qquad (1.59)$$

where $dt = n \, ds/c$ denotes the infinitesimal time interval taken by light in moving an infinitesimal distance ds in a medium with refractive index n and the space metric is denoted by g_{ij}. The geodesic properties of light propagation are investigated with the *vacuum* metric g_{ij} or the *medium-modified* metric $\overline{g}_{ij} = n^2 \, g_{ij}$.

1.4.3.1 *Vacuum-metric case*

We begin with the vacuum-metric case and consider the light-curvature equation (1.50). First, we define the vacuum-metric tensor $g_{ij} = \mathbf{e}_i \cdot \mathbf{e}_j$ in terms of the basis vectors $(\mathbf{e}_1, \mathbf{e}_2, \mathbf{e}_3)$, so that the ray velocity is

$$\frac{d\mathbf{x}}{ds} = \frac{dx^i}{ds}\, \mathbf{e}_i.$$

Second, using the definition for the Christoffel symbol (1.18) and the relations

$$\frac{d\mathbf{e}_j}{ds} \equiv \Gamma^i_{jk}\, \frac{dx^k}{ds}\, \mathbf{e}_i,$$

we find

$$\frac{d\widehat{\mathbf{k}}}{ds} \equiv \frac{d^2\mathbf{x}}{ds^2} = \frac{d^2x^i}{ds^2}\, \mathbf{e}_i + \frac{dx^i}{ds}\frac{d\mathbf{e}_i}{ds} = \left(\frac{d^2x^i}{ds^2} + \Gamma^i_{jk}\, \frac{dx^j}{ds}\frac{dx^k}{ds} \right) \mathbf{e}_i.$$

By combining these relations, light-curvature equation (1.50) becomes

$$\frac{d^2x^i}{ds^2} + \Gamma^i_{jk}\, \frac{dx^j}{ds}\frac{dx^k}{ds} = \left(g^{ij} - \frac{dx^i}{ds}\frac{dx^j}{ds} \right) \frac{\partial \ln n}{\partial x^j}. \tag{1.60}$$

This equation shows that the path of a light ray departs from a vacuum geodesic line as a result of a refractive-index gradient projected along the tensor

$$h^{ij} \equiv g^{ij} - \frac{dx^i}{ds}\frac{dx^j}{ds}$$

which, by construction, is perpendicular to the ray velocity $d\mathbf{x}/ds$ (i.e., $h^{ij}\, dx_j/ds = 0$).

1.4.3.2 *Medium-metric case*

Next, we investigate the geodesic propagation of a light ray associated with the medium-modified (conformal) metric $\overline{g}_{ij} = n^2\, g_{ij}$, where $c^2 dt^2 = n^2 ds^2 = \overline{g}_{ij}\, dx^i dx^j$. The derivation follows a variational formulation similar to that found in Sec. 1.1.3. Hence, the first-order variation $\delta T[\mathbf{x}]$ is expressed as

$$\delta T[\mathbf{x}] = \int_{t_A}^{t_B} \left[\frac{d^2x^i}{dt^2} + \overline{\Gamma}^i_{jk}\, \frac{dx^j}{dt}\frac{dx^k}{dt} \right] \overline{g}_{i\ell}\, \delta x^\ell\, \frac{dt}{c^2}, \tag{1.61}$$

where the medium-modified Christoffel symbol $\overline{\Gamma}^i_{jk}$ includes the effects of the gradient in the refractive index $n(\mathbf{x})$. We, therefore, find that the light path $\mathbf{x}(t)$ is a solution of the geodesic equation

$$\frac{d^2x^i}{dt^2} + \overline{\Gamma}^i_{jk}\, \frac{dx^j}{dt}\frac{dx^k}{dt} = 0, \tag{1.62}$$

which is also the path of least time for which $\delta \mathcal{T}[\mathbf{x}] = 0$. When using Cartesian coordinates (where $\overline{g}_{ij} = n^2 \delta_{ij}$ and $\overline{g}^{ij} = n^{-2}\delta^{ij}$), for example, the medium-modified Christoffel symbol

$$\overline{\Gamma}^i_{jk} = \delta^i_j \, \partial_k \ln n + \delta^i_k \, \partial_j \ln n - \delta_{jk} \, \delta^{i\ell} \, \partial_\ell \ln n$$

is expressed in terms of gradient-components of the logarithm of the refraction index n.

1.4.3.3 *Jacobi equation for light propagation*

Lastly, we point out that the Jacobi equation for the deviation $\boldsymbol{\xi}(\sigma) = \overline{\mathbf{x}}(\sigma) - \mathbf{x}(\sigma)$ between two rays that satisfy the Euler-Fermat ray equation (1.48) can be obtained from the Jacobi function

$$J(\boldsymbol{\xi}, d\boldsymbol{\xi}/d\sigma) \equiv \frac{1}{2} \left[\frac{d^2}{d\epsilon^2} \left(n(\mathbf{x} + \epsilon \boldsymbol{\xi}) \left| \frac{d\mathbf{x}}{d\sigma} + \epsilon \frac{d\boldsymbol{\xi}}{d\sigma} \right| \right) \right]_{\epsilon=0} \qquad (1.63)$$

$$\equiv \frac{n}{2\Lambda^3} \left| \frac{d\boldsymbol{\xi}}{d\sigma} \times \frac{d\mathbf{x}}{d\sigma} \right|^2 + \frac{\boldsymbol{\xi} \cdot \nabla n}{\Lambda} \frac{d\boldsymbol{\xi}}{d\sigma} \cdot \frac{d\mathbf{x}}{d\sigma} + \frac{\Lambda}{2} \boldsymbol{\xi}\boldsymbol{\xi} : \nabla\nabla n,$$

where the Euler-Fermat ray equation (1.48) was taken into account and the exact σ-derivative, which cancels out upon integration, is omitted. Hence, the Jacobi equation describing light-ray deviation is expressed as the Jacobi-Euler-Fermat equation

$$\frac{d}{d\sigma} \left(\frac{\partial J}{\partial (d\boldsymbol{\xi}/d\sigma)} \right) = \frac{\partial J}{\partial \boldsymbol{\xi}},$$

which yields

$$\frac{d}{d\sigma} \left[\frac{n}{\Lambda^3} \frac{d\mathbf{x}}{d\sigma} \times \left(\frac{d\boldsymbol{\xi}}{d\sigma} \times \frac{d\mathbf{x}}{d\sigma} \right) \right] = \Lambda \, \boldsymbol{\xi} \cdot \nabla\nabla n \cdot \left(\mathbf{I} - \frac{1}{\Lambda^2} \frac{d\mathbf{x}}{d\sigma} \frac{d\mathbf{x}}{d\sigma} \right) \qquad (1.64)$$

$$+ \left[\frac{d\boldsymbol{\xi}}{d\sigma} - (\boldsymbol{\xi} \cdot \nabla \ln n) \frac{d\mathbf{x}}{d\sigma} \right] \times \left(\frac{\nabla n}{\Lambda} \times \frac{d\mathbf{x}}{d\sigma} \right).$$

The Jacobi equation (1.64) describes the property of nearby rays to converge or diverge in a nonuniform refractive medium. Note, here, that the terms involving $\Lambda^{-1}\nabla n \times d\mathbf{x}/d\sigma$ in Eq. (1.64) can be written in terms of the Euler-Fermat ray equation (1.48) as

$$\frac{\nabla n}{\Lambda} \times \frac{d\mathbf{x}}{d\sigma} = \frac{1}{\Lambda^2} \frac{d}{d\sigma} \left(\frac{n}{\Lambda} \frac{d\mathbf{x}}{d\sigma} \right) \times \frac{d\mathbf{x}}{d\sigma} = \frac{n}{\Lambda^3} \left(\frac{d^2\mathbf{x}}{d\sigma^2} \times \frac{d\mathbf{x}}{d\sigma} \right),$$

which, thus, involve the Frenet-Serret ray curvature.

1.4.4   *Wavefront Representation*

The complementary picture of rays propagating in a nonuniform medium was proposed by Christiaan Huygens (1629-1695) in terms of the picture of propagating wavefronts. Here, a wavefront is defined as the surface that is locally perpendicular to a ray. Hence, the index of refraction itself (for an isotropic medium) can be written as

$$n = |\nabla S| = \frac{ck}{\omega} \ \text{ or } \ \nabla S = n \frac{d\mathbf{x}}{ds} = \frac{c\mathbf{k}}{\omega}, \tag{1.65}$$

where S is called the *eikonal* function (i.e., a wavefront is defined by the surface $S =$ constant; see Fig. 1.8). To show that this definition is consistent with Eq. (1.50), we easily check that

$$\frac{d}{ds}\left(n \frac{d\mathbf{x}}{ds} \right) = \frac{d\nabla S}{ds} = \frac{d\mathbf{x}}{ds} \cdot \nabla \nabla S = \frac{1}{n} \nabla S \cdot \nabla \nabla S$$
$$= \frac{1}{2n} \nabla |\nabla S|^2 = \frac{1}{2n} \nabla n^2 = \nabla n.$$

This definition, therefore, implies that the wavevector $\mathbf{k}$ is curl-free:

$$\nabla \times \mathbf{k} = \nabla \times \nabla \left(\frac{\omega}{c} S \right) \equiv 0, \tag{1.66}$$

where we used the fact that the wave frequency ω is unchanged by refraction. Hence, we find that $\nabla \times \widehat{\mathbf{k}} = \widehat{\mathbf{k}} \times \nabla \ln n$, from which we obtain the light-curvature equation (1.50). Note also that because $\mathbf{k}$ is curl-free, we easily apply Stokes' Theorem to find that the closed contour integral $\oint_{\partial A} \mathbf{k} \cdot d\mathbf{x} = 0$ along the boundary ∂A of an open surface A vanishes, i.e., the path integral $\int \mathbf{k} \cdot d\mathbf{x}$ is path-independent.

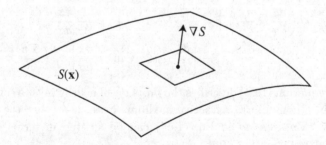

Fig. 1.8   Wavefront surface.

Lastly, in the absence of sources and sinks, the light energy flux entering a finite volume bounded by a closed surface is equal to the light energy flux leaving the volume and, thus, the intensity of light I satisfies the conservation law

$$0 = \nabla \cdot (I \nabla S) = I \nabla^2 S + \nabla S \cdot \nabla I. \qquad (1.67)$$

Using the definition $\nabla S \cdot \nabla \equiv n \, \partial/\partial s$, we find the intensity *evolution* equation

$$\frac{\partial \ln I}{\partial s} = - n^{-1} \nabla^2 S,$$

whose solution is expressed as

$$I = I_0 \exp\left(- \int_0^s \nabla^2 S \, \frac{d\sigma}{n}\right), \qquad (1.68)$$

where I_0 is the light intensity at position $s = 0$ along a ray. This equation, therefore, determines whether light intensity increases ($\nabla^2 S < 0$) or decreases ($\nabla^2 S > 0$) along a ray depending on the sign of $\nabla^2 S$. In a refractive medium with spherical symmetry, with $S'(r) = n(r)$ and $\hat{k} = \hat{r}$, the conservation law (1.67) becomes

$$0 = \frac{1}{r^2} \frac{d}{dr} \left(r^2 I \, n\right),$$

which implies that the light intensity satisfies the inverse-square law: $I(r) n(r) \, r^2 = I_0 n_0 \, r_0^2$.

1.5   Problems

1. Find Euler's First and Second equations following the extremization of the integral

$$\mathcal{F}[y] = \int_a^b F(y, y', y'') \, dx.$$

State whether an additional set of boundary conditions for $\delta y'(a)$ and $\delta y'(b)$ are necessary.

2. Find the curve joining two points (x_1, y_1) and (x_2, y_2) that yields a surface of revolution (about the x-axis) of minimum area by minimizing the integral

$$A[y] = \int_{x_1}^{x_2} y \sqrt{1 + (y')^2} \, dx.$$

3. Use the Jacobi equation (1.10) to obtain Eq. (1.13) for $\delta^2 \mathcal{F}$.

4. This problem deals with finding the equation for geodesics on a cone represented by $z = \rho \cot \alpha$, for which the infinitesimal length element ds is defined as

$$ds^2 = d\rho^2 + \rho^2 \, d\phi^2 + dz^2 = \left[\rho^2 + \csc^2\alpha \, (\rho')^2 \right] d\phi^2.$$

(a) Show that Euler's Second equation for $\rho(\phi)$ can be written as

$$\frac{\rho^2 \, \sin\alpha}{\sqrt{\rho^2 \, \sin^2\alpha + (\rho')^2}} = \rho_0,$$

where $\rho_0 \equiv \rho(\phi_0)$ and $\rho'(\phi_0) \equiv 0$.

(b) Solve Euler's Second equation for $\rho(\phi)$ and show that the equation for geodesics on a cone is

$$\rho(\phi) = \rho_0 \, \sec\left[\sin^2\alpha \, (\phi - \phi_0) \right].$$

5. Show that the time required for a particle to move without friction from the point (x_0, y_0) parametrized by the angle θ_0 to the minimum point $(\pi a, 2a)$ of the cycloid solution of the brachistochrone problem is $\pi \sqrt{a/g}$.

6. A thin rope of mass m (and uniform density) is attached to two vertical poles of height H separated by a horizontal distance D; the coordinates of the pole tops are set at $(\pm D/2, H)$. If the length L of the rope is greater than D, it will sag under the action of gravity and its lowest point (at its midpoint) will be at a height $y(x = 0) = y_0$. The shape of the rope, subject to the boundary conditions $y(\pm D/2) = H$, is obtained by minimizing the gravitational potential energy of the rope expressed in terms of the functional

$$\mathcal{U}[y] = \int_{D/2}^{-D/2} mg \, y \, \sqrt{1 + (y')^2} \, dx.$$

Show that the extremal curve $y(x)$ (known as the *catenary* curve) for this problem is

$$y(x) = c \, \cosh\left(\frac{x - b}{c} \right),$$

where $b = 0$ and $c = y_0$.

7. Show that the parametric solution given by Eqs. (1.44)-(1.45) for the linear refractive medium can be expressed as Eq. (1.46).

8. A light ray travels in a medium with refractive index

$$n(y) = n_0 \exp(-\beta y),$$

where n_0 is the refractive index at $y = 0$ and β is a positive constant.

(a) Use the results of the Principle of Least Time contained in the Notes distributed in class to show that the path of the light ray is expressed as

$$y(x; \beta) = \frac{1}{\beta} \ln \left[\frac{\cos(\beta x - \varphi_0)}{\cos \varphi_0} \right], \qquad (1.69)$$

where the light ray is initially traveling upwards from $(x, y) = (0, 0)$ at an angle φ_0.

(b) Using the appropriate mathematical techniques, show that we recover the expected result $\lim_{\beta \to 0} y(x; \beta) = (\tan \varphi_0) x$ from Eq. (1.69).

(c) The light ray reaches a maximum height $\overline{y}$ at $x = \overline{x}(\beta)$, where $y'(\overline{x}; \beta) = 0$. Find expressions for $\overline{x}$ and $\overline{y}(\beta) = y(\overline{x}; \beta)$.

9. Consider the path associated with the index of refraction $n(y) = H/y$, where the height H is a constant and $0 < y < H\alpha^{-1} \equiv R$ to ensure that, according to Eq. (1.40), $n(y) > \alpha$. Show that the light path has the simple semi-circular form:

$$(R - x)^2 + y^2 = R^2 \quad \to \quad y(x) = \sqrt{x(2R - x)}.$$

10.* Using the parametric solutions (1.44)-(1.45) of the optical path in a linear refractive medium, calculate the Frenet-Serret curvature coefficient

$$\kappa(\varphi) = \frac{|\mathbf{r}''(\varphi) \times \mathbf{r}'(\varphi)|}{|\mathbf{r}'(\varphi)|^3},$$

and show that it is equal to $|\hat{\mathbf{k}} \times \nabla \ln n|$.

11. Assuming that the refractive index $n(z)$ in a nonuniform medium is a function of z only, derive the Euler-Fermat equations (1.50) for the components (α, β, γ) of the unit vector $\widehat{k} = \alpha\widehat{x} + \beta\widehat{y} + \gamma\widehat{z}$.

12. In Fig. 1.7, show that the angle $\varphi(\theta)$ defined from the conservation law (1.55) is expressed as

$$\varphi(\theta) = \arcsin\left[\frac{1 + e\cos 2\theta}{\sqrt{1 + e^2 + 2e\cos 2\theta}}\right],$$

so that $\varphi = \pi/2$ at $\theta = 0$ and $\pi/2$, as expected for an ellipse.

13. Find the light-path trajectory $r(\theta)$ for a spherically-symmetric medium with index of refraction $n(r) = n_0\,(b/r)^2$, where b is an arbitrary constant and $n_0 = n(b)$.

14.* Derive the Jacobi equation (1.64) for two-dimensional light propagation in a nonuniform medium with index of refraction $n(y)$; *Hint*: choose $\sigma = x$. Compare your Jacobi equation with that obtained from Eq. (1.10).

15.* Lagrange showed in 1760 that a surface $z(x, y)$ has minimal area if it satisfies the partial differential equation

$$\left(1 + q^2\right)\frac{\partial^2 z}{\partial x^2} + \left(1 + p^2\right)\frac{\partial^2 z}{\partial y^2} - 2\,p\,q\,\frac{\partial^2 z}{\partial x\,\partial y} = 0,$$

where $(p, q) \equiv (\partial z/\partial x,\ \partial z/\partial y)$.

(a) Derive this equation by minimizing the surface integral

$$I[z] = \int\int \sqrt{1 + p^2 + q^2}\ dx\,dy.$$

(b) Show that the surface $z(x, y) = \cosh^{-1}(\sqrt{x^2 + y^2})$ has minimal area.

16. (a) Show that the optical length followed by a light ray along the path APB in Fig. 1.9 is $L(\theta) = 2\sqrt{2}\,R\,\cos(\theta/2)$, where R is the radius of the circle.

(b) Show that the optical length $L(\theta)$ has a *maximum* for $\theta = 0$.

17. We now consider light propagation in axially-symmetric cylindrical geometry, where the index of refraction $n(\rho)$ is a function of the cylindrical radius ρ (measured from the z-axis). If we use the z-coordinate as the ray

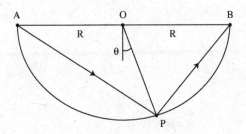

Fig. 1.9   Problem 16.

parameter, Fermat's Principle of Least Time (1.37) becomes

$$\delta \int_a^b n(\rho)\, \Lambda(\rho, \dot\rho, \dot\theta)\, dz \;\equiv\; \delta \int_a^b n(\rho)\, \sqrt{1 + \dot\rho^2 + \rho^2\dot\theta^2}\; dz \;=\; 0,$$

where $\dot\rho = d\rho/dz$ and $\dot\theta = d\theta/dz$. Note that the integrand $F \equiv n\Lambda$ is independent of z and θ and therefore

$$N \equiv F - \dot\rho\, \frac{\partial F}{\partial \dot\rho} - \dot\theta\, \frac{\partial F}{\partial \dot\theta}, \tag{1.70}$$

$$R \equiv \frac{\partial F}{\partial \dot\theta}, \tag{1.71}$$

are constants along the light path (i.e., $dN/dz = 0 = dR/dz$).

(a) Using the conservation law (1.71), show that, by solving for $\dot\theta$ as a function of ρ and $\dot\rho$, we obtain

$$\Lambda(\rho, \dot\rho) \;\equiv\; n(\rho)\, \rho\, \sqrt{\frac{1 + \dot\rho^2}{n^2(\rho)\rho^2 - R^2}}. \tag{1.72}$$

(b) Using the conservation law (1.70), with Eq. (1.72), obtain the integral solution

$$z(\rho) \;\equiv\; z_a + \int_a^\rho \frac{N\sigma\, d\sigma}{\sqrt{\sigma^2[n^2(\sigma) - N^2] - R^2}},$$

which can then be inverted to obtain $\rho(z; N, R)$.

Chapter 2

Lagrangian Mechanics

Newtonian mechanics discusses the dynamics of particles in terms of (vector) forces acting on them. In contrast, the investigation of particle dynamics within Lagrangian mechanics uses the concepts of kinetic and potential (scalar) energies. The difference may seem academic until we realize that it is the Lagrangian method which generalizes to physical theories that lie well beyond the classical dynamics of particles.

In this Chapter, we present four principles by which single-particle dynamics may be derived. The reader is referred to Refs. [4, 11, 18] for comments regarding the history of the Principles of Least Action of Maupertuis, Jacobi, and Hamilton as well as Refs. [8–10] for some additional comments concerning recent developments. The primary focus of this Chapter will be applications of Hamilton's Principle on which Lagrangian Mechanics is based.

2.1 Maupertuis-Jacobi Principle of Least Action

The publication of Fermat's Principle of Least Time in 1657 generated an intense controversy between Fermat and disciples of René Descartes (1596-1650) regarding whether light travels slower (Fermat) or faster (Descartes) in a dense medium as compared to free space.

In 1744, Pierre Louis Moreau de Maupertuis (1698-1759) stated (without proof) that, in analogy with Fermat's Principle of Least Time for light, a particle of mass m under the influence of a force $\mathbf{F} = -\nabla U$ moves along a path that satisfies the Principle of *Least Action*: $\delta S = 0$, where the action integral is defined as

$$S[\mathbf{x}] = \int \mathbf{p} \cdot d\mathbf{x} = \int mv \, ds. \tag{2.1}$$

Here $v = ds/dt$ denotes the magnitude of particle velocity, which can also be expressed as

$$v(s) = \sqrt{(2/m)[E - U(s)]}, \qquad (2.2)$$

with the particle's kinetic energy $K = mv^2/2 = E - U$ written in terms of its total energy E and its potential energy $U(s)$.

2.1.1 *Maupertuis' Principle*

In 1744, Euler proved Maupertuis' Principle of Least Action $\delta S = 0$ for particle motion in the (x, y)-plane as follows [18]. For this purpose, we use the Frenet-Serret curvature formula for the path $y(x)$, where we define the tangent unit vector $\widehat{t}$ and the principal normal unit vector $\widehat{n}$ as

$$\widehat{t} = \frac{d\mathbf{x}}{ds} = \frac{\widehat{x} + y'\,\widehat{y}}{\sqrt{1 + (y')^2}} \quad \text{and} \quad \widehat{n} = \frac{y'\,\widehat{x} - \widehat{y}}{\sqrt{1 + (y')^2}} \equiv (-\widehat{z}) \times \widehat{t}, \qquad (2.3)$$

where $y' = dy/dx$ and $ds = dx\sqrt{1 + (y')^2}$. The Frenet-Serret formula for the curvature of a two-dimensional curve is

$$\frac{d\widehat{t}}{ds} = \frac{|y''|\,\widehat{n}}{[1 + (y')^2]^{3/2}} = \kappa\,\widehat{n},$$

where the instantaneous radius of curvature ρ is defined as $\rho = \kappa^{-1}$.

By using Newton's Second Law of Motion $\mathbf{F} = -\nabla U$ and the energy conservation law $(\nabla K = -\nabla U)$, we find the relation

$$\mathbf{F} = mv\left(\frac{dv}{ds}\,\widehat{t} + v\,\frac{d\widehat{t}}{ds}\right) = \widehat{t}\,(\widehat{t} \cdot \nabla K) + mv^2\,\kappa\,\widehat{n}$$

$$= \nabla K = mv\,\nabla v \qquad (2.4)$$

between the unit vectors $\widehat{t}$ and $\widehat{n}$ associated with the path, the Frenet-Serret curvature κ, and the kinetic energy $K = \frac{1}{2}\,mv^2(x, y)$ of the particle. Note that Eq. (2.4) can be re-written as

$$\frac{d\widehat{t}}{ds} = \widehat{t} \times (\nabla \ln v \times \widehat{t}), \qquad (2.5)$$

which highlights a deep connection with Eq. (1.50) derived from Fermat's Principle of Least Time, where the index of refraction n is now replaced by the speed $v = \sqrt{(2/m)[E - U(s)]}$. Lastly, we point out that the type of dissipationless forces considered in Eq. (2.4) involves *active* forces (defined as forces that do work), as opposed to *passive* forces (defined as forces that do no work, such as constraint forces).

Next, the action integral (2.1) is expressed as

$$S = \int m\,v(x,y)\,\sqrt{1+(y')^2}\,dx = \int F(y,y';x)\,dx, \qquad (2.6)$$

so that

$$\frac{\partial F}{\partial y'} = \frac{mv\,y'}{\sqrt{1+(y')^2}} \quad \text{and} \quad \frac{\partial F}{\partial y} = m\sqrt{1+(y')^2}\,\frac{\partial v}{\partial y}.$$

We thus obtain the Euler-Maupertuis equation

$$\frac{m\,v\,y''}{[1+(y')^2]^{3/2}} = \frac{m}{\sqrt{1+(y')^2}}\,\frac{\partial v}{\partial y} - \frac{m\,y'}{\sqrt{1+(y')^2}}\,\frac{\partial v}{\partial x} \equiv m\,\hat{n}\cdot\nabla v, \qquad (2.7)$$

which can also be expressed as $mv\,\kappa = m\,\hat{n}\cdot\nabla v$. Using the relation $\mathbf{F} = \nabla K$ and the Frenet-Serret formulas (2.3), the Maupertuis-Euler equation (2.7) becomes

$$mv^2\,\kappa = \mathbf{F}\cdot\hat{n},$$

from which we recover Newton's Second Law (2.4).

2.1.2 *Jacobi's Principle*

Jacobi emphasized the connection between Fermat's Principle of Least Time (1.35) and Maupertuis' Principle of Least Action (2.1) by introducing a different form of the Principle of Least Action $\delta S = 0$, where Jacobi's action integral is

$$S[\mathbf{x}] = \int \sqrt{2m\,(E-U)}\,ds = 2\int K\,dt, \qquad (2.8)$$

where particle momentum is written as $p = \sqrt{2m\,(E-U)}$. To obtain the second expression of the action integral (2.8), Jacobi made use of the fact that, by introducing a path parameter τ such that $v = ds/dt = s'/t'$ (where a prime denotes a τ-derivative), we find

$$K = \frac{m\,(s')^2}{2\,(t')^2} = E - U,$$

so that $2K\,t' = s'\,p$, and the second form of Jacobi's action integral results. Next, Jacobi used the Principle of Least Action (2.8) to establish the geometric foundations of particle mechanics. Here, the Euler-Jacobi equation resulting from Jacobi's Principle of Least Action is expressed as

$$\frac{d}{ds}\left(\sqrt{E-U}\,\frac{d\mathbf{x}}{ds}\right) = \nabla\sqrt{E-U}, \qquad (2.9)$$

which is identical to the Euler-Fermat equation (1.49), with the index of refraction n substituted with $\sqrt{E-U}$.

Note that the connection between Fermat's Principle of Least Time and Maupertuis-Jacobi's Principle of Least Action yields the relation

$$n = \gamma \, |\mathbf{p}|, \tag{2.10}$$

where γ is a constant (see Table 3.1). This connection was later used by Prince Louis Victor Pierre Raymond de Broglie (1892-1987) to establish the relation $|\mathbf{p}| = \hbar|\mathbf{k}| = n\,(\hbar\omega/c)$ between the momentum of a particle and its wavenumber $|\mathbf{k}| = 2\pi/\lambda = n\,\omega/c$. Using de Broglie's relation $\mathbf{p} = \hbar\mathbf{k}$ between the particle momentum $\mathbf{p}$ and the wave vector $\mathbf{k}$, we note that

$$\frac{|\mathbf{p}|^2}{2m} = \frac{\hbar^2|\mathbf{k}|^2}{2m} \equiv \frac{\hbar^2}{2m}\,|\nabla\Theta|^2 = E - U, \tag{2.11}$$

where Θ is the dimensionless (eikonal) phase. The time-independent Hamilton-Jacobi equation

$$E = \frac{|\nabla \mathcal{S}_E|^2}{2m} + U \tag{2.12}$$

is obtained from Eq. (2.11) by using the relation $\mathcal{S}_E \equiv \hbar\,\Theta$ between Hamilton's principal function $\mathcal{S}_E$ (at constant energy E) and the eikonal phase Θ (see Chap. 3 for additional details). Further historical comments concerning the variational derivation of Schroedinger's equation is discussed by Dugas [4], Lanczos [11], and Yourgrau and Mandelstam [18], as well as Refs. [9, 10].

2.2 d'Alembert's Principle

So far, the Maupertuis-Jacobi principles (2.1) and (2.8) make use of the length variable s as the orbit parameter to describe particle motion. We now turn our attention to two principles that will provide a clear path toward the ultimate action principle called Hamilton's Principle, from which equations of motion are derived in terms of generalized spatial coordinates in *configuration* space.

First, within the context of Newtonian mechanics, we distinguish between two classes of forces, depending on whether a force is able to do work or not. In the first class, an *active* force $\mathbf{F}_w$ is involved in performing infinitesimal work $dW = \mathbf{F}_w \cdot d\mathbf{x}$ evaluated along the infinitesimal displacement $d\mathbf{x}$; the class of active forces includes conservative (e.g., gravity) and nonconservative (e.g., friction) forces. In the second class, a *passive* force (labeled $\mathbf{F}_0$) is defined as a force not involved in performing work, which includes constraint forces such as normal and tension forces. Here, the infinitesimal work performed by a passive force is $\mathbf{F}_0 \cdot d\mathbf{x} = 0$ because the infinitesimal displacement $d\mathbf{x}$ is required to satisfy the constraints.

2.2.1 *Principle of Virtual Work*

The Principle of Virtual Work is one of the oldest principles in Physics, which may find its origin in the work of Aristotle (384-322 B.C.) on the static equilibrium of levers [4]. The Principle of Virtual Work was finally written in its current form in 1717 by Jean (Johann) Bernoulli and states that a system composed of N particles is in static equilibrium if the virtual work

$$\delta W = \sum_{i=1}^{N} \mathbf{F}_i \cdot \delta \mathbf{x}^i = 0 \tag{2.13}$$

vanishes for all virtual displacements $(\delta \mathbf{x}^1, \ldots, \delta \mathbf{x}^N)$ that satisfy physical constraints.

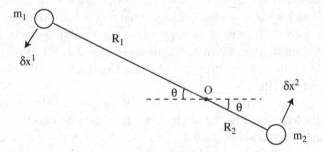

Fig. 2.1   Static equilibrium of a lever.

As an application of the Principle of Virtual Work (2.13), we consider the static equilibrium of a lever (see Fig. 2.1) composed of two masses m_1 and m_2 placed on a massless rod at distances R_1 and R_2, respectively, from the fulcrum point O. Here, the only active forces acting on the masses are due to gravity: $\mathbf{F}_i = -m_i\, g\, \widehat{\mathbf{y}}$, and the position vectors of m_1 and m_2 are

$$\mathbf{r}_1 = R_1\left(-\cos\theta\,\widehat{\mathbf{x}} + \sin\theta\,\widehat{\mathbf{y}}\right) \quad \text{and} \quad \mathbf{r}_2 = R_2\left(\cos\theta\,\widehat{\mathbf{x}} - \sin\theta\,\widehat{\mathbf{y}}\right),$$

respectively (see Fig. 2.1). By using the virtual displacements

$$\delta \mathbf{x}^i = \epsilon\widehat{\mathbf{z}} \times \mathbf{r}_i \tag{2.14}$$

(where ϵ is an infinitesimal angular displacement and the axis of rotation is directed along the z-axis), the Principle of Virtual Work (2.13) yields the following condition for static equilibrium:

$$0 = \epsilon \cos\theta \, (m_1 g \, R_1 - m_2 g \, R_2) \quad \rightarrow \quad m_1 R_1 = m_2 R_2.$$

Note that, although the static equilibrium of the lever is based on the concept of *torque* (moment of force) equilibrium, the Principle of Virtual Work shows that all static equilibria are encompassed by the Principle.

2.2.2 *Lagrange's Equations from d'Alembert's Principle*

It was Jean Le Rond d'Alembert (1717-1783) who generalized the Principle of Virtual Work (in 1742) by including the *accelerating* force $- m_i \, \ddot{\mathbf{x}}^i$ in the Principle of Virtual Work (2.13):

$$\sum_{i=1}^{N} \left(\mathbf{F}_i - m_i \frac{d^2 \mathbf{x}^i}{dt^2} \right) \cdot \delta \mathbf{x}^i = 0 \qquad (2.15)$$

so that the equations of dynamics could be obtained from Eq. (2.15). Hence, d'Alembert's Principle, in effect, states that the work done by all active forces acting in a system is algebraically equal to the work done by all the acceleration forces. Note that, in contrast to the variational principles of Classical Mechanics (e.g., Fermat, Maupertuis, and Jacobi), d'Alembert's Principle (2.15) and Gauss' Principle of Least Constraint (see problem 11) are *constraint* principles.

As a simple application of d'Alembert's Principle (2.15), we return to the lever problem (see Fig. 2.1), where we now assume $m_2 R_2 > m_1 R_1$. Here, the particle accelerations are

$$\ddot{\mathbf{x}}^i = -(\dot{\theta})^2 \, \mathbf{r}_i - \ddot{\theta} \, \widehat{\mathbf{z}} \times \mathbf{r}_i,$$

so that, with Eq. (2.14), d'Alembert's Principle (2.15) yields

$$0 = \epsilon \left[(m_1 g \, R_1 - m_2 g \, R_2) \cos\theta + \left(m_1 R_1^2 + m_2 R_2^2 \right) \ddot{\theta} \right].$$

Hence, according to d'Alembert's Principle, the angular acceleration $\ddot{\theta}$ of the unbalanced lever is

$$\ddot{\theta} = \frac{g \cos\theta}{I} \, (m_2 R_2 - m_1 R_1),$$

where $I = m_1 R_1^2 + m_2 R_2^2$ denotes the moment of inertia of the lever as it rotates about the fulcrum point O. Thus, we see that rotational dynamics associated with unbalanced torques can be described in terms of d'Alembert's Principle. (See Chap. 7 for additional details concerning rotational dynamics.)

The most historically significant application of d'Alembert's Principle (2.15), however, came from Lagrange who transformed it as follows. Consider, for simplicity, the following infinitesimal-work identity

$$0 = \left(\mathbf{F} - m \frac{d^2\mathbf{x}}{dt^2} \right) \cdot \delta\mathbf{x}$$

$$= \delta W - \frac{d}{dt}\left(m \frac{d\mathbf{x}}{dt} \cdot \delta\mathbf{x} \right) + m \frac{d\mathbf{x}}{dt} \cdot \frac{d\,\delta\mathbf{x}}{dt}, \tag{2.16}$$

where $\mathbf{F}$ represents an active force applied to a particle of mass m so that $\delta W = \mathbf{F} \cdot \delta\mathbf{x}$ denotes the virtual work calculated along the virtual displacement $\delta\mathbf{x}$. We note that if the position vector $\mathbf{x}(q_1, ..., q_k; t)$ is a time-dependent function of k *generalized* coordinates, then we find

$$\delta\mathbf{x} = \sum_{i=1}^{k} \frac{\partial\mathbf{x}}{\partial q_i} \delta q_i,$$

and

$$\mathbf{v} = \frac{d\mathbf{x}}{dt} = \frac{\partial\mathbf{x}}{\partial t} + \sum_{i=1}^{k} \frac{\partial\mathbf{x}}{\partial q_i} \dot{q}_i. \tag{2.17}$$

Next, we introduce the variation of the kinetic energy $K = mv^2/2$:

$$\delta K = m \frac{d\mathbf{x}}{dt} \cdot \frac{d\,\delta\mathbf{x}}{dt} = \sum_{i} \delta q_i \frac{\partial K}{\partial q_i},$$

since the virtual variation operator δ (introduced by Lagrange) commutes with the time derivative d/dt, and we introduce the generalized force

$$Q^i \equiv \mathbf{F} \cdot \frac{\partial\mathbf{x}}{\partial q_i},$$

so that $\delta W = \sum_i Q^i \delta q_i$. We shall also use the identity

$$m \frac{d^2\mathbf{x}}{dt^2} \cdot \frac{\partial\mathbf{x}}{\partial q_i} = \frac{d}{dt}\left(m\mathbf{v} \cdot \frac{\partial\mathbf{x}}{\partial q_i} \right) - m\mathbf{v} \cdot \frac{d}{dt}\left(\frac{\partial\mathbf{x}}{\partial q_i} \right),$$

with

$$\frac{d}{dt}\left(\frac{\partial\mathbf{x}}{\partial q_i} \right) \equiv \frac{\partial^2\mathbf{x}}{\partial t\,\partial q_i} + \dot{q}_j \frac{\partial^2\mathbf{x}}{\partial q_j\,\partial q_i} \equiv \frac{\partial\mathbf{v}}{\partial q_i},$$

and

$$\frac{\partial\mathbf{v}}{\partial \dot{q}_i} = \frac{\partial\mathbf{x}}{\partial q_i},$$

which both follow from Eq. (2.17). Our first result derived from d'Alembert's Principle (2.16) is now expressed in terms of the generalized coordinates $(q_1, ..., q_k)$ as

$$0 = \sum_i \delta q_i \left[\frac{d}{dt} \left(\frac{\partial K}{\partial \dot{q}_i} \right) - \frac{\partial K}{\partial q_i} - Q^i \right].$$

Since this relation must hold for any variation δq_i $(i = 1, ..., k)$, we obtain the d'Alembert-Lagrange equation

$$\frac{d}{dt} \left(\frac{\partial K}{\partial \dot{q}_i} \right) - \frac{\partial K}{\partial q_i} = Q^i, \tag{2.18}$$

where we note that the generalized force Q^i is associated with any active (conservative or nonconservative) force **F**. Hence, for a *conservative* active force derivable from a single potential energy U (i.e., $\mathbf{F} = -\nabla U$), the ith-component of the generalized force is $Q^i = -\partial U/\partial q_i$, and the d'Alembert-Lagrange equation (2.18) becomes

$$\frac{d}{dt} \left(\frac{\partial K}{\partial \dot{q}_i} \right) - \frac{\partial K}{\partial q_i} = -\frac{\partial U}{\partial q_i}. \tag{2.19}$$

We shall return to this important equation (see Eq. (2.28)).

Our second result based on d'Alembert's Principle (2.16), now expressed as

$$\delta K + \delta W = \frac{d}{dt} \left(m \frac{d\mathbf{x}}{dt} \cdot \delta \mathbf{x} \right), \tag{2.20}$$

is obtained as follows. For a *conservative* active force derivable from a single potential energy U (i.e., $\mathbf{F} = -\nabla U$), the virtual work is $\delta W = -\delta U$, so that time integration of Eq. (2.20) yields an important principle known as Hamilton's Principle

$$\int_{t_1}^{t_2} \left(\delta K - \delta U \right) dt \equiv \delta \int_{t_1}^{t_2} L \, dt = 0, \tag{2.21}$$

where $\delta \mathbf{x}$ vanishes at $t = t_1$ and t_2 and the function $L = K - U$, obtained by subtracting the potential energy U from the kinetic energy K, is known as the Lagrangian function of the system. Note that the Maupertuis-Jacobi Principle (2.8) leads to Hamilton's Principle (2.21) if we use the identity $2K \equiv (K - U) + E$ and use the variation operator δ_E at constant total energy E.

2.3 Hamilton's Principle

2.3.1 *Constraint Forces*

To illustrate Hamilton's Principle (2.21), we consider a pendulum composed of an object of mass m and a massless string of constant length ℓ in a constant gravitational field with acceleration g. We first investigate the motion of the pendulum as a dynamical problem in two dimensions with a single constraint (i.e., constant length) and later reduce this problem to a single dimension by carefully choosing a single generalized coordinate.

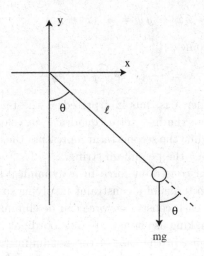

Fig. 2.2 The two-dimensional pendulum problem.

Using Cartesian coordinates (x, y) for the pendulum mass shown in Fig. 2.2, the kinetic energy is $K = \frac{1}{2}m(\dot{x}^2 + \dot{y}^2)$ and the gravitational potential energy is $U = mg\,y$, where the length of the pendulum string ℓ is *constrained* to be constant (i.e., $\ell^2 = x^2 + y^2$). Hence, the constrained action integral is expressed as

$$\mathcal{A}_\lambda[\mathbf{x}] = \int \left[\frac{1}{2}m\left(\dot{x}^2 + \dot{y}^2\right) - mg\,y + \lambda\left(\ell - \sqrt{x^2+y^2}\right) \right] dt$$

$$\equiv \int F(\mathbf{x}, \dot{\mathbf{x}}; \lambda)\, dt,$$

where λ represents a Lagrange multiplier used to enforce the constant-length constraint (see Sec. 1.2.1). By definition, we find $\partial F/\partial \lambda = 0$ for all

x. Euler's equations for x and y, respectively, are

$$m\ddot{x} = -\lambda \frac{x}{\ell} \quad \text{and} \quad m\ddot{y} = -mg - \lambda \frac{y}{\ell}. \tag{2.22}$$

Using the constant-length constraint, the Lagrange multiplier λ is constructed from Eqs. (2.22) and expressed as

$$\lambda \equiv -\frac{m}{\ell} \left[gy + (x\ddot{x} + y\ddot{y}) \right]. \tag{2.23}$$

Next, using the second time derivative of the constant-length constraint $\ell^2 = x^2 + y^2$, we obtain

$$x\ddot{x} + y\ddot{y} = -\left(\dot{x}^2 + \dot{y}^2\right),$$

so that Eq. (2.23) becomes

$$\lambda = \frac{m}{\ell}\left(\dot{x}^2 + \dot{y}^2\right) - mg\frac{y}{\ell}.$$

The Lagrange multiplier λ is thus interpreted as the tension force in the pendulum string, where the first term (quadratic in velocities) represents the centrifugal force while the second term represents the component of the gravitational force along the pendulum string.

It turns out that a constraint force in a dynamical system can most often be represented in terms of a constraint involving spatial coordinates. We shall now see that each constraint force can be eliminated from the dynamical problem by making use of new spatial coordinates that enforce the constraint. For example, in the case of the pendulum problem discussed above, we note that the constant-length constraint can be enforced by expressing the Cartesian coordinates $x = \ell \sin\theta$ and $y = -\ell \cos\theta$ in terms of the angle θ (see Fig. 2.2).

2.3.2 *Generalized Coordinates in Configuration Space*

The *configuration* space of a mechanical system with constraints evolving in n-dimensional space, with spatial coordinates $\mathbf{x} = (x^1, x^2, ..., x^n)$, can sometimes be described in terms of *generalized* coordinates $\mathbf{q} = (q^1, q^2, ..., q^k)$ in a k-dimensional *configuration* space, with $k \leq n$. Each generalized coordinate q^i is said to describe motion along a *degree of freedom* of the mechanical system.

For example, consider a mechanical system composed of two particles (see Fig. 2.3), with masses (m_1, m_2) and three-dimensional coordinate positions $(\mathbf{x}_1, \mathbf{x}_2)$, tied together with a massless rigid rod (so that the distance

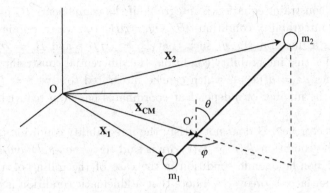

Fig. 2.3 Configuration space.

$|\mathbf{x}_1 - \mathbf{x}_2|$ is constant). The configuration of this two-particle system (in six-dimensional space) can be' described in terms of the coordinates

$$\mathbf{x}_{CM} \equiv \frac{\sum_i m_i \, \mathbf{x}_i}{\sum_i m_i} = \frac{m_1 \, \mathbf{x}_1 + m_2 \, \mathbf{x}_2}{m_1 + m_2} \tag{2.24}$$

of the center-of-mass (CM) in the Laboratory frame (O) and the orientation of the rod in the CM frame (O') expressed in terms of the two angles (θ, φ). Hence, as a result of the existence of a single constraint ($\ell = |\mathbf{x}_1 - \mathbf{x}_2|$), the generalized coordinates for this system are $(\mathbf{x}_{CM}; \theta, \varphi)$ and we have reduced the number of coordinates needed to describe the state of the system from six to five. Each generalized coordinate is said to describe dynamics along a *degree of freedom* of the mechanical system; for example, in the case of the two-particle system discussed above, the generalized coordinates $\mathbf{x}_{CM}$ describe an arbitrary translation of the center-of-mass while the generalized coordinates (θ, φ) describe an arbitrary rotation about the center-of-mass.

Constraints are found to be of two different types referred to as *holonomic* and *nonholonomic* constraints [11]. For example, the differential (kinematic) constraint equation $dq(\mathbf{r}) = \mathbf{B}(\mathbf{r}) \cdot d\mathbf{r}$ is said to be *holonomic* (or integrable) if the vector field $\mathbf{B}$ satisfies the integrability condition

$$\nabla \times \mathbf{B} = 0. \tag{2.25}$$

If this condition is satisfied, the function $q(\mathbf{r})$ can be explicitly constructed and, thus, the number of independent coordinates can be reduced by one. For example, consider the differential constraint equation $dz = B_x(x, y) \, dx + B_y(x, y) \, dy$, where an infinitesimal change in the x and y coordinates produce an infinitesimal change in the z coordinate. This differential constraint equation is integrable if the components B_x and B_y satisfy the integrability condition $\partial B_x/\partial y = \partial B_y/\partial x$, which implies that there exists a function $f(x, y)$ such that $B_x = \partial f/\partial x$ and $B_y = \partial f/\partial y$. Hence, under this integrability condition, the differential constraint equation becomes $dz = df(x, y)$, which can be integrated to give $z = f(x, y)$ and, thus, the number of independent coordinates has been reduced from 3 to 2.

If the vector field **B** does not satisfy the integrability condition (2.25), however, the condition $dq(\mathbf{r}) = \mathbf{B}(\mathbf{r}) \cdot d\mathbf{r}$ is said to be *non-holonomic*. An example of non-holonomic condition is the case of the rolling of a solid body on a surface. Moreover, we note that a kinematic condition is called *rheonomic* if it is time-dependent, otherwise it is called *scleronomic*.

In summary, the presence of holonomic constraints can always be treated by the introduction of generalized coordinates. The treatment of nonholonomic constraints, on the other hand, requires the addition of constraint forces on the right side of Lagrange's equation (2.18), which falls outside the scope of this course.

2.3.3 *Constrained Motion on a Surface*

As an example of motion under an holonomic constraint, we consider the general problem associated with the motion of a particle constrained to move on a surface described by the relation $F(x, y, z) = 0$. First, since the velocity $d\mathbf{x}/dt$ of the particle along its trajectory must be perpendicular to the gradient ∇F (i.e., tangent to the surface $F = 0$), the displacement $d\mathbf{x}$ is required to satisfy the constraint condition

$$d\mathbf{x} \cdot \nabla F = 0. \tag{2.26}$$

Next, any point $\mathbf{x}$ on the surface $F(x, y, z) = 0$ may be parameterized by two *surface* coordinates (u, v) such that

$$\frac{\partial \mathbf{x}}{\partial u}(u, v) \cdot \nabla F = 0 = \frac{\partial \mathbf{x}}{\partial v}(u, v) \cdot \nabla F.$$

Hence, we may write an expression for $d\mathbf{x}$ that satisfies Eq. (2.26) as

$$d\mathbf{x} = \frac{\partial \mathbf{x}}{\partial u}\, du + \frac{\partial \mathbf{x}}{\partial v}\, dv \ \text{ and } \ \frac{\partial \mathbf{x}}{\partial u} \times \frac{\partial \mathbf{x}}{\partial v} = \mathcal{J}\, \nabla F,$$

where the function $\mathcal{J}$ depends on the surface coordinates (u, v). It is thus quite clear that the surface coordinates (u, v) are the generalized coordinates for this constrained motion.

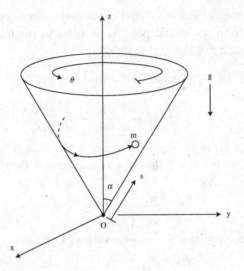

Fig. 2.4   Motion on the surface of a cone.

For example, we consider the motion of a particle constrained to move on the surface of a cone of apex angle α (see Fig. 2.4). Here, the constraint is expressed as $F(x, y, z) = \sqrt{x^2 + y^2} - z \tan \alpha = 0$ with $\nabla F = \widehat{\rho} - \tan \alpha \, \widehat{z}$, where $\rho^2 = x^2 + y^2$ and $\widehat{\rho} = (x/\rho)\widehat{x} + (y/\rho)\widehat{y}$. The surface coordinates can be chosen to be the polar angle θ and the function

$$s(x, y, z) = \sqrt{x^2 + y^2 + z^2} \equiv \sqrt{\rho^2 + z^2},$$

which measures the distance from the apex of the cone (defining the origin). Hence, we find

$$\frac{\partial \mathbf{x}}{\partial \theta} = \rho \, \widehat{\theta} = \rho \widehat{z} \times \widehat{\rho} \quad \text{and} \quad \frac{\partial \mathbf{x}}{\partial s} = \sin \alpha \, \widehat{\rho} + \cos \alpha \, \widehat{z} = \widehat{s},$$

with

$$\frac{\partial \mathbf{x}}{\partial \theta} \times \frac{\partial \mathbf{x}}{\partial s} = \rho \cos \alpha \, \nabla F$$

and $\mathcal{J} = \rho \cos \alpha$. We shall return to this example in Sec. 2.5.4.

2.3.4   *Euler-Lagrange Equations*

Hamilton's principle (sometimes called THE Principle of Least Action) is expressed in terms of a function $L(\mathbf{q}, \dot{\mathbf{q}}; t)$ known as the *Lagrangian*, which appears in the *action* integral

$$S[\mathbf{q}] = \int_{t_i}^{t_f} L(\mathbf{q}, \dot{\mathbf{q}}; t) \, dt, \qquad (2.27)$$

where the action integral is a functional of the generalized coordinates $\mathbf{q}(t)$, providing a path from the initial point $\mathbf{q}_i = \mathbf{q}(t_i)$ to the final point $\mathbf{q}_f = \mathbf{q}(t_f)$. The stationarity of the action integral

$$0 = \delta\mathcal{S}[\mathbf{q}; \delta\mathbf{q}] = \left(\frac{d}{d\epsilon} \mathcal{S}[\mathbf{q} + \epsilon\,\delta\mathbf{q}] \right)_{\epsilon=0}$$

$$= \int_{t_i}^{t_f} \delta\mathbf{q} \cdot \left[\frac{\partial L}{\partial \mathbf{q}} - \frac{d}{dt}\left(\frac{\partial L}{\partial \dot{\mathbf{q}}} \right) \right]\,dt,$$

where the variation $\delta\mathbf{q}$ is assumed to vanish at the integration boundaries ($\delta\mathbf{q}_i = 0 = \delta\mathbf{q}_f$), yields the *Euler-Lagrange* equation for the generalized coordinate q^j ($j = 1, ..., k$):

$$\frac{d}{dt}\left(\frac{\partial L}{\partial \dot{q}^j} \right) = \frac{\partial L}{\partial q^j}. \tag{2.28}$$

The Lagrangian also satisfies the Euler's Second Equation:

$$\frac{d}{dt}\left(L - \dot{\mathbf{q}} \cdot \frac{\partial L}{\partial \dot{\mathbf{q}}} \right) = \frac{\partial L}{\partial t}, \tag{2.29}$$

and thus, for time-independent Lagrangian systems ($\partial L/\partial t = 0$), we find that $L - \dot{\mathbf{q}} \cdot \partial L/\partial \dot{\mathbf{q}}$ is a conserved quantity whose interpretation will be discussed shortly.

Note that, according to d'Alembert's Principle (2.19), the form of the Lagrangian function $L(\mathbf{r}, \dot{\mathbf{r}}; t)$ is dictated by our requirement that Newton's Second Law $m\,\ddot{\mathbf{r}} = -\nabla U(\mathbf{r}, t)$, which describes the motion of a particle of mass m in a nonuniform (possibly time-dependent) potential $U(\mathbf{r}, t)$, be written in the Euler-Lagrange form (2.28). One easily obtains the form

$$L(\mathbf{r}, \dot{\mathbf{r}}; t) = \frac{m}{2} |\dot{\mathbf{r}}|^2 - U(\mathbf{r}, t), \tag{2.30}$$

for the Lagrangian of a particle of mass m, which is simply the kinetic energy of the particle **minus** its potential energy. The minus sign in Eq. (2.30) is important; not only does this form give us the correct equations of motion but, without the minus sign, energy would not be conserved. In fact, we note that Jacobi's action integral (2.8) can also be written as $A = \int [(K - U) + E]\,dt$, using the energy conservation law $E = K + U$; hence, energy conservation is the important connection between the Principles of Least Action of Maupertuis-Jacobi and Euler-Lagrange.

2.3.5 *Four-step Lagrangian Method*

For a simple mechanical system, the Lagrangian function is obtained by computing the kinetic energy of the system and its potential energy and then construct Eq. (2.30). The construction of a Lagrangian function for a system of N particles, therefore, proceeds in four steps as follows.

• **Step I.** Define k generalized coordinates $\mathbf{q}(t) = (q^1(t), ..., q^k(t))$ that represent the instantaneous *configuration* of the mechanical system of N particles at time t. Hence, for each particle (labeled $a = 1, ..., N$), the Cartesian-coordinate position vector

$$\mathbf{r}_a \equiv \mathbf{r}_a(\mathbf{q}; t) \tag{2.31}$$

is expressed as an explicit function of the generalized coordinates.

• **Step II.** For each particle, use the position vector (2.31) to construct the velocity

$$\mathbf{v}_a(\mathbf{q}, \dot{\mathbf{q}}; t) = \frac{\partial \mathbf{r}_a}{\partial t} + \sum_{j=1}^{k} \dot{q}^j \frac{\partial \mathbf{r}_a}{\partial q^j}. \tag{2.32}$$

• **Step III.** From the position (2.31) and velocity (2.32) of each particle, construct the kinetic energy

$$K(\mathbf{q}, \dot{\mathbf{q}}; t) = \sum_a \frac{m_a}{2} \, |\mathbf{v}_a(\mathbf{q}, \dot{\mathbf{q}}; t)|^2$$

and the potential energy

$$U(\mathbf{q}; t) = \sum_a U(\mathbf{r}_a(\mathbf{q}; t), t)$$

for the system and combine them to obtain the Lagrangian

$$L(\mathbf{q}, \dot{\mathbf{q}}; t) = K(\mathbf{q}, \dot{\mathbf{q}}; t) - U(\mathbf{q}; t). \tag{2.33}$$

• **Step IV.** From the Lagrangian (2.33), the Euler-Lagrange equations (2.28) are derived for each generalized coordinate q^j:

$$\sum_a \frac{d}{dt} \left(\frac{\partial \mathbf{r}_a}{\partial q^j} \cdot m_a \mathbf{v}_a \right) = \sum_a \left(m_a \mathbf{v}_a \cdot \frac{\partial \mathbf{v}_a}{\partial q^j} - \frac{\partial \mathbf{r}_a}{\partial q^j} \cdot \nabla_a U \right), \tag{2.34}$$

where we have used the identity $\partial \mathbf{v}_a / \partial \dot{q}^j = \partial \mathbf{r}_a / \partial q^j$.

2.3.6 *Lagrangian Mechanics in Curvilinear Coordinates**

Jacobi was the first to investigate the relation between particle dynamics and Riemannian geometry. The Euler-Lagrange equation (2.34) can be framed within the context of Riemannian geometry as follows. The kinetic energy of a single particle of mass m, with generalized coordinates $\mathbf{q} = (q^1, ..., q^k)$, is expressed as

$$K = \frac{m}{2} \, |\mathbf{v}|^2 = \frac{m}{2} \frac{\partial \mathbf{r}}{\partial q^i} \cdot \frac{\partial \mathbf{r}}{\partial q^j} \, \dot{q}^i \, \dot{q}^j \equiv \frac{m}{2} \, g_{ij} \, \dot{q}^i \, \dot{q}^j,$$

where g_{ij} denotes the metric tensor on configuration space. When the particle moves in a potential $U(\mathbf{q})$, the Euler-Lagrange equation (2.34) becomes

$$\frac{d}{dt}\left(m\,g_{ij}\,\dot{q}^j\right) = m\,g_{ij}\,\ddot{q}^j + \frac{m}{2}\left(\frac{\partial g_{ij}}{\partial q^k} + \frac{\partial g_{ik}}{\partial q^j}\right)\dot{q}^j\,\dot{q}^k$$

$$= \frac{m}{2}\frac{\partial g_{jk}}{\partial q^i}\,\dot{q}^j\,\dot{q}^k - \frac{\partial U}{\partial q^i},$$

or

$$m\,g_{ij}\left(\frac{d^2 q^j}{dt^2} + \Gamma^j_{k\ell}\frac{dq^k}{dt}\frac{dq^\ell}{dt}\right) = -\frac{\partial U}{\partial q^i}, \qquad (2.35)$$

where the Christoffel symbol (1.18) is defined as

$$\Gamma^j_{k\ell} \equiv \frac{g^{ij}}{2}\left(\frac{\partial g_{ik}}{\partial q^\ell} + \frac{\partial g_{i\ell}}{\partial q^k} - \frac{\partial g_{k\ell}}{\partial q^i}\right).$$

Thus, the concepts associated with Riemannian geometry that appear extensively in the theory of General Relativity have natural antecedents in classical Lagrangian mechanics.

2.4 Lagrangian Mechanics in Configuration Space

In this Section, we explore the Lagrangian formulation of several mechanical systems listed here in order of increasing complexity. As we proceed with our examples, we should realize how the Lagrangian formulation maintains its relative simplicity compared to the application of the more familiar Newton's method (Isaac Newton, 1643-1727) associated with the vectorial decomposition of forces. Here, all constraint forces are eliminated in terms of generalized coordinates and all active conservative forces are expressed in terms of gradients of suitable potential energies.

2.4.1 *Example I: Pendulum*

As a first example, we reconsider the pendulum (see Sec. 2.3.1) composed of an object of mass m and a massless string of constant length ℓ in a constant gravitational field with acceleration g. Although the motion of the pendulum is two-dimensional, a single generalized coordinate is needed to describe the configuration of the pendulum: the angle θ measured from the negative y-axis (see Fig. 2.2). Here, the position of the object is given as

$$x(\theta) = \ell\sin\theta \quad \text{and} \quad y(\theta) = -\ell\cos\theta,$$

with associated velocity components

$$\dot{x}(\theta, \dot{\theta}) = \ell\dot{\theta}\,\cos\theta \quad \text{and} \quad \dot{y}(\theta, \dot{\theta}) = \ell\dot{\theta}\,\sin\theta.$$

Hence, the kinetic energy of the pendulum is

$$K = \frac{m}{2}\left(\dot{x}^2 + \dot{y}^2\right) = \frac{m}{2}\,\ell^2\dot{\theta}^2,$$

and choosing the zero potential energy point when $\theta = 0$, the gravitational potential energy is

$$U = mg\ell\,(1 - \cos\theta).$$

The Lagrangian $L = K - U$ is, therefore, written as

$$L(\theta, \dot{\theta}) = \frac{m}{2}\,\ell^2\dot{\theta}^2 - mg\ell\,(1 - \cos\theta),$$

and the Euler-Lagrange equation for θ is

$$\frac{\partial L}{\partial \dot{\theta}} = m\ell^2\dot{\theta} \;\rightarrow\; \frac{d}{dt}\left(\frac{\partial L}{\partial \dot{\theta}}\right) = m\ell^2\ddot{\theta}$$

$$\frac{\partial L}{\partial \theta} = -mg\ell\,\sin\theta$$

or

$$\ddot{\theta} + \frac{g}{\ell}\,\sin\theta = 0. \tag{2.36}$$

The pendulum problem (2.36) is solved in the next Chapter through the use of the Energy method. Note that, whereas the tension in the pendulum string must be considered explicitly in the Newtonian method, the string tension is replaced by the constraint $\ell = $ constant in the Lagrangian method.

2.4.2   *Example II: Bead on a Rotating Hoop*

As a second example, we consider a bead of mass m sliding freely on a hoop of radius R rotating with angular velocity Ω in a constant gravitational field with acceleration g (see Fig. 2.5). Here, since the bead on the rotating hoop effectively moves on the surface of a sphere of radius R, we use the generalized coordinates given by the two angles θ (measured from the negative z-axis) and φ (measured from the positive x-axis), where $\dot{\varphi} = \Omega$ is used as an additional constraint. The position of the bead is given in terms of Cartesian coordinates as

$$x(\theta, t) = R\,\sin\theta\,\cos(\varphi_0 + \Omega t),$$
$$y(\theta, t) = R\,\sin\theta\,\sin(\varphi_0 + \Omega t),$$
$$z(\theta, t) = -R\,\cos\theta,$$

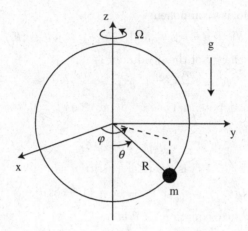

Fig. 2.5   Generalized coordinates for the bead-on-a-rotating-hoop problem.

where $\varphi(t) = \varphi_0 + \Omega\,t$, and its associated Cartesian velocity components are

$$\dot{x}(\theta, \dot{\theta}; t) = R\left(\dot{\theta}\,\cos\theta\,\cos\varphi \;-\; \Omega\,\sin\theta\,\sin\varphi\right),$$

$$\dot{y}(\theta, \dot{\theta}; t) = R\left(\dot{\theta}\,\cos\theta\,\sin\varphi \;+\; \Omega\,\sin\theta\,\cos\varphi\right),$$

$$\dot{z}(\theta, \dot{\theta}; t) = R\,\dot{\theta}\,\sin\theta,$$

so that the kinetic energy of the bead is

$$K(\theta, \dot{\theta}) \;=\; \frac{m}{2}\,|\mathbf{v}|^2 \;=\; \frac{m\,R^2}{2}\left(\dot{\theta}^2 \;+\; \Omega^2\,\sin^2\theta\right).$$

The gravitational potential energy is

$$U(\theta) \;=\; mgR\,(1\,-\,\cos\theta),$$

where the zero-potential energy point is chosen at $\theta = 0$ (see Fig. 2.5).

The Lagrangian $L = K - U$ is, therefore, written as

$$L(\theta, \dot{\theta}) \;=\; \frac{m\,R^2}{2}\left(\dot{\theta}^2 \;+\; \Omega^2\,\sin^2\theta\right) \;-\; mgR\,(1 - \cos\theta),$$

and the Euler-Lagrange equation for θ is

$$\frac{\partial L}{\partial \dot{\theta}} \;=\; mR^2\,\dot{\theta} \;\;\rightarrow\;\; \frac{d}{dt}\left(\frac{\partial L}{\partial \dot{\theta}}\right) = mR^2\,\ddot{\theta}$$

$$\frac{\partial L}{\partial \theta} = -\,mgR\,\sin\theta$$

$$+\; mR^2\Omega^2\,\cos\theta\,\sin\theta$$

or

$$\ddot{\theta} + \sin\theta \left(\frac{g}{R} - \Omega^2 \cos\theta \right) = 0$$

Note that the support (constraint) force provided by the hoop (necessary in the Newtonian method) is now replaced by the constraint $R = $ constant in the Lagrangian method. Furthermore, although the motion intrinsically takes place on the surface of a sphere of radius R, the azimuthal motion is completely determined by the equation $\varphi(t) = \varphi_0 + \Omega t$ and, thus, the motion of the bead takes place in one dimension.

Lastly, we note that this equation displays *bifurcation* behavior, which is investigated in Chap. 8. For $\Omega^2 < g/R$, the equilibrium point $\theta = 0$ is stable while, for $\Omega^2 > g/R$, the equilibrium point $\theta = 0$ is now unstable and the new equilibrium point $\theta = \arccos(g/\Omega^2 R)$ is stable.

2.4.3 *Example III: Rotating Pendulum*

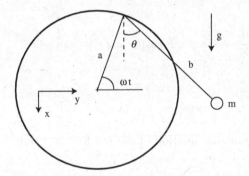

Fig. 2.6 Generalized coordinates for the rotating-pendulum problem.

As a third example, we consider a pendulum of mass m and length b attached to the edge of a disk of radius a rotating at angular velocity ω in a constant gravitational field with acceleration g. Placing the origin at the center of the disk, the coordinates of the pendulum mass are

$$(x,\, y) = a\,(-\sin\omega t,\ \cos\omega t) + b\,(\cos\theta,\ \sin\theta),$$

so that the velocity components are

$$(\dot{x},\, \dot{y}) = -a\omega\,(\cos\omega t,\ \sin\omega t) + b\dot{\theta}\,(-\sin\theta,\ \cos\theta),$$

and the squared velocity is

$$v^2 = a^2\omega^2 + b^2\dot{\theta}^2 + 2\,ab\,\omega\,\dot{\theta}\,\sin(\theta - \omega t).$$

Setting the zero potential energy at $x = 0$, the gravitational potential energy is

$$U = -mgx = mga \sin \omega t - mgb \cos \theta.$$

The Lagrangian $L = K - U$ is, therefore, written as

$$L(\theta, \dot{\theta}; t) = \frac{m}{2} \left[a^2 \omega^2 + b^2 \dot{\theta}^2 + 2ab \omega \dot{\theta} \sin(\theta - \omega t) \right]$$
$$- mga \sin \omega t + mgb \cos \theta, \qquad (2.37)$$

and the Euler-Lagrange equation for θ is

$$\frac{\partial L}{\partial \dot{\theta}} = mb^2 \dot{\theta} + mab \omega \sin(\theta - \omega t) \quad \rightarrow$$
$$\frac{d}{dt} \left(\frac{\partial L}{\partial \dot{\theta}} \right) = mb^2 \ddot{\theta} + mab \omega (\dot{\theta} - \omega) \cos(\theta - \omega t)$$

and

$$\frac{\partial L}{\partial \theta} = mab \omega \dot{\theta} \cos(\theta - \omega t) - mgb \sin \theta$$

or

$$\ddot{\theta} + \frac{g}{b} \sin \theta - \frac{a}{b} \omega^2 \cos(\theta - \omega t) = 0$$

We recover the standard equation of motion for the pendulum when a or ω vanish.

Note that the terms

$$\frac{m}{2} a^2 \omega^2 - mga \sin \omega t$$

in the Lagrangian (2.37) play no role in determining the dynamics of the system. In fact, as can easily be shown (see Sec. 2.5), a Lagrangian L is always defined up to an exact time derivative, i.e., the Lagrangians L and $L' = L - df/dt$, where $f(\mathbf{q}, t)$ is an arbitrary function, lead to the same Euler-Lagrange equations. In the present case,

$$f(t) = [(m/2) a^2 \omega^2] t + (mga/\omega) \cos \omega t$$

and thus this term can be omitted from the Lagrangian (2.37) without changing the equations of motion.

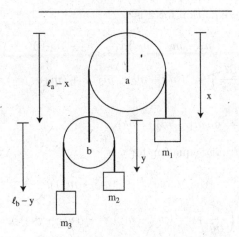

Fig. 2.7   Generalized coordinates for the compound-Atwood problem.

2.4.4   *Example IV: Compound Atwood Machine*

As a fourth (and penultimate) example, we consider a compound Atwood machine (see Fig. 2.7) composed three masses (labeled m_1, m_2, and m_3) attached by two massless ropes through two massless pulleys in a constant gravitational field with acceleration g.

The two generalized coordinates for this system are the distance x of mass m_1 from the top of the first pulley and the distance y of mass m_2 from the top of the second pulley; here, the lengths ℓ_a and ℓ_b are constants. The coordinates and velocities of the three masses m_1, m_2, and m_3 are

$$x_1 = x \rightarrow v_1 = \dot{x},$$

$$x_2 = \ell_a - x + y \rightarrow v_2 = \dot{y} - \dot{x},$$

$$x_3 = \ell_a - x + \ell_b - y \rightarrow v_3 = -\dot{x} - \dot{y},$$

respectively, so that the total kinetic energy is

$$K = \frac{m_1}{2}\dot{x}^2 + \frac{m_2}{2}(\dot{y} - \dot{x})^2 + \frac{m_3}{2}(\dot{x} + \dot{y})^2.$$

Placing the zero potential energy at the top of the first pulley, the total gravitational potential energy, on the other hand, can be written as

$$U = -gx(m_1 - m_2 - m_3) - gy(m_2 - m_3),$$

where constant terms were omitted. The Lagrangian $L = K - U$ is, therefore, written as

$$L(x, \dot{x}, y, \dot{y}) = \frac{m_1}{2}\dot{x}^2 + \frac{m_2}{2}(\dot{x} - \dot{y})^2 + \frac{m_3}{2}(\dot{x} + \dot{y})^2$$
$$+ gx(m_1 - m_2 - m_3) + gy(m_2 - m_3).$$

The Euler-Lagrange equation for x is

$$\frac{\partial L}{\partial \dot{x}} = (m_1 + m_2 + m_3)\, \dot{x} + (m_3 - m_2)\, \dot{y} \ \rightarrow$$

$$\frac{d}{dt}\left(\frac{\partial L}{\partial \dot{x}}\right) = (m_1 + m_2 + m_3)\, \ddot{x} + (m_3 - m_2)\, \ddot{y}$$

$$\frac{\partial L}{\partial x} = g\, (m_1 - m_2 - m_3)$$

while the Euler-Lagrange equation for y is

$$\frac{\partial L}{\partial \dot{y}} = (m_3 - m_2)\, \dot{x} + (m_2 + m_3)\, \dot{y} \ \rightarrow$$

$$\frac{d}{dt}\left(\frac{\partial L}{\partial \dot{y}}\right) = (m_3 - m_2)\, \ddot{x} + (m_2 + m_3)\, \ddot{y}$$

$$\frac{\partial L}{\partial y} = g\, (m_2 - m_3).$$

We combine these two Euler-Lagrange equations

$$(m_1 + m_2 + m_3)\, \ddot{x} + (m_3 - m_2)\, \ddot{y} = g\, (m_1 - m_2 - m_3), \qquad (2.38)$$

$$(m_3 - m_2)\, \ddot{x} + (m_2 + m_3)\, \ddot{y} = g\, (m_2 - m_3), \qquad (2.39)$$

to describe the dynamical evolution of the compound Atwood machine. This set of equations can, in fact, be solved explicitly as

$$\ddot{x} = g\left(\frac{m_1\, m_+ - (m_+^2 - m_-^2)}{m_1\, m_+ + (m_+^2 - m_-^2)}\right)$$

and

$$\ddot{y} = g\left(\frac{2\, m_1\, m_-}{m_1\, m_+ + (m_+^2 - m_-^2)}\right),$$

where $m_\pm = m_2 \pm m_3$. Note also that, by using the energy conservation law $E = K + U$ it can be shown that the position z of the center of mass of the mechanical system (as measured from the top of the first pulley) satisfies the relation

$$Mg\,(z - z_0) = \frac{m_1}{2}\, \dot{x}^2 + \frac{m_2}{2}\, (\dot{y} - \dot{x})^2 + \frac{m_3}{2}\, (\dot{x} + \dot{y})^2 > 0, \qquad (2.40)$$

where $M = (m_1 + m_2 + m_3)$ denotes the total mass of the system and we have assumed that the system starts from rest (with its center of mass located at z_0). This important relation tells us that, as the masses start to move ($\dot{x} \neq 0$ and $\dot{y} \neq 0$), the center of mass must fall: $z > z_0$.

Before proceeding to our next (and last) example, we introduce a convenient technique (henceforth known as *Frozen Degrees of Freedom*) for checking on the physical accuracy of any set of coupled Euler-Lagrange equations. Hence, for the Euler-Lagrange equation (2.38), we may freeze the degree of freedom associated with the y-coordinate (i.e., we set $\dot{y} = 0 = \ddot{y}$ or $m_- = 0$) to obtain $\ddot{x} = g\,(m_1 - m_+)/(m_1 + m_+)$, in agreement with the analysis of a *simple* Atwood machine composed of a mass m_1 on one side and a mass $m_+ = m_2 + m_3$ on the other side. Likewise, for the Euler-Lagrange equation (2.39), we may freeze the degree of freedom associated with the x-coordinate (i.e., we set $\dot{x} = 0 = \ddot{x}$ or $m_1 m_+ = m_+^2 - m_-^2$) to obtain $\ddot{y} = g\,(m_-/m_+)$, again in agreement with the analysis of a *simple* Atwood machine.

2.4.5 *Example V: Pendulum with Oscillating Fulcrum*

As a fifth and final example, we consider the case of a pendulum of mass m and length ℓ attached to a massless block which is attached to a fixed wall by a massless spring of constant k. Here, we assume that the massless block moves without friction on a set of rails (see problem 5). We use the two

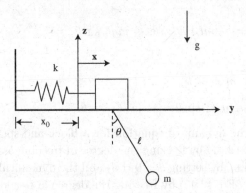

Fig. 2.8 Generalized coordinates for the oscillating-pendulum problem.

generalized coordinates x and θ shown in Fig. 2.8 and write the Cartesian coordinates (y, z) of the pendulum mass as $y = x + \ell \sin\theta$ and $z = -\ell \cos\theta$, with its associated velocity components $\dot{y} = \dot{x} + \ell\dot{\theta} \cos\theta$ and $\dot{z} = \ell\dot{\theta} \sin\theta$. The kinetic energy of the pendulum is thus

$$K = \frac{m}{2}\left(\dot{y}^2 + \dot{z}^2\right) = \frac{m}{2}\left(\dot{x}^2 + \ell^2\dot{\theta}^2 + 2\ell \cos\theta\, \dot{x}\dot{\theta}\right).$$

The potential energy $U = U_k + U_g$ has two terms: one term $U_k = \frac{1}{2} k x^2$ associated with displacement of the spring away from its equilibrium position and one term $U_g = mgz$ associated with gravity. Hence, the Lagrangian for this system is

$$L(x, \theta, \dot{x}, \dot{\theta}) = \frac{m}{2} \left(\dot{x}^2 + \ell^2 \dot{\theta}^2 + 2\ell \cos\theta \, \dot{x}\dot{\theta} \right) - \frac{k}{2} x^2 + mg\ell \cos\theta.$$

The Euler-Lagrange equation for x is

$$\frac{\partial L}{\partial \dot{x}} = m \left(\dot{x} + \ell \cos\theta \, \dot{\theta} \right) \rightarrow$$

$$\frac{d}{dt} \left(\frac{\partial L}{\partial \dot{x}} \right) = m \ddot{x} + m\ell \left(\ddot{\theta} \cos\theta - \dot{\theta}^2 \sin\theta \right)$$

$$\frac{\partial L}{\partial x} = - k x$$

while the Euler-Lagrange equation for θ is

$$\frac{\partial L}{\partial \dot{\theta}} = m\ell \left(\ell \dot{\theta} + \dot{x} \cos\theta \right) \rightarrow$$

$$\frac{d}{dt} \left(\frac{\partial L}{\partial \dot{\theta}} \right) = m\ell^2 \ddot{\theta} + m\ell \left(\ddot{x} \cos\theta - \dot{x}\dot{\theta} \sin\theta \right)$$

$$\frac{\partial L}{\partial \theta} = - m\ell \dot{x}\dot{\theta} \sin\theta - mg\ell \sin\theta$$

or

$$m \ddot{x} + k x = m\ell \left(\dot{\theta}^2 \sin\theta - \ddot{\theta} \cos\theta \right), \qquad (2.41)$$

$$\ddot{\theta} + (g/\ell) \sin\theta = - (\ddot{x}/\ell) \cos\theta. \qquad (2.42)$$

Here, we recover the dynamical equation for a block-and-spring harmonic oscillator from Eq. (2.41) by freezing the degree of freedom associated with the θ-coordinate (i.e., by setting $\dot{\theta} = 0 = \ddot{\theta}$) and the dynamical equation for the pendulum from Eq. (2.42) by freezing the degree of freedom associated with the x-coordinate (i.e., by setting $\dot{x} = 0 = \ddot{x}$). It is easy to see from this last example how powerful and yet simple the Lagrangian method is compared to the Newtonian method.

2.5 Symmetries and Conservation Laws

We are sometimes faced with a Lagrangian function that is either independent of time, independent of a linear spatial coordinate, or independent of an angular spatial coordinate. The Noether theorem (Amalie Emmy Noether, 1882-1935) states that *for each symmetry of the Lagrangian there corresponds a conservation law (and vice versa).* When the Lagrangian L is invariant under a time translation, a space translation, or a spatial rotation, the conservation law involves energy, linear momentum, or angular momentum, respectively.

We begin our discussion with a general expression for the variation δL of the Lagrangian $L(\mathbf{q}, \dot{\mathbf{q}}, t)$:

$$\delta L = \delta \mathbf{q} \cdot \left[\frac{\partial L}{\partial \mathbf{q}} - \frac{d}{dt} \left(\frac{\partial L}{\partial \dot{\mathbf{q}}} \right) \right] + \frac{d}{dt} \left(\delta \mathbf{q} \cdot \frac{\partial L}{\partial \dot{\mathbf{q}}} \right),$$

obtained after re-arranging the term $\delta \dot{\mathbf{q}} \cdot \partial L / \partial \dot{\mathbf{q}}$. Next, we make use of the Euler-Lagrange equations for $\mathbf{q}$ (which enables us to drop the term $\delta \mathbf{q} \cdot [\cdots]$) and we find

$$\delta L = \frac{d}{dt} \left(\delta \mathbf{q} \cdot \frac{\partial L}{\partial \dot{\mathbf{q}}} \right). \tag{2.43}$$

Lastly, the variation δL can only be generated by a time translation δt, since

$$0 = \delta \int L \, dt = \int \left[\left(\delta L + \delta t \, \frac{\partial L}{\partial t} \right) dt + L \, d\delta t \right]$$
$$= \int \left[\delta L - \delta t \left(\frac{dL}{dt} - \frac{\partial L}{\partial t} \right) \right] dt$$

so that

$$\delta L = \delta t \left(\frac{dL}{dt} - \frac{\partial L}{\partial t} \right)$$

By combining this expression with Eq. (2.43), we find

$$\delta t \left(\frac{dL}{dt} - \frac{\partial L}{\partial t} \right) = \frac{d}{dt} \left(\delta \mathbf{q} \cdot \frac{\partial L}{\partial \dot{\mathbf{q}}} \right), \tag{2.44}$$

which we, henceforth, refer to as the Noether equation for finite-dimensional mechanical systems (see Chap. 9 for the infinite-dimensional case).

We now apply Noether's Theorem, based on the Noether equation (2.44), to investigate the connection between symmetries of the Lagrangian with conservation laws.

2.5.1 *Energy Conservation Law*

First, we consider time translations, $t \to t + \delta t$ and $\delta \mathbf{q} = \dot{\mathbf{q}}\, \delta t$, so that the Noether equation (2.44) becomes Euler's Second Equation

$$-\frac{\partial L}{\partial t} = \frac{d}{dt} \left(\dot{\mathbf{q}} \cdot \frac{\partial L}{\partial \dot{\mathbf{q}}} - L \right).$$

Noether's Theorem states that if the Lagrangian is invariant under time translations, i.e., $\partial L / \partial t = 0$, then energy is conserved, $dE/dt = 0$, where

$$E = \dot{\mathbf{q}} \cdot \frac{\partial L}{\partial \dot{\mathbf{q}}} - L$$

defines the energy invariant.

2.5.2 *Momentum Conservation Laws*

Next, we consider invariance under spatial translations, $\mathbf{q} \to \mathbf{q} + \boldsymbol{\epsilon}$ (where $\delta \mathbf{q} = \boldsymbol{\epsilon}$ denotes a constant infinitesimal displacement and $\delta t = 0$), so that the Noether equation (2.44) yields the *linear* momentum conservation law

$$0 = \frac{d}{dt} \left(\frac{\partial L}{\partial \dot{\mathbf{q}}} \right) = \frac{d\mathbf{P}}{dt},$$

where $\mathbf{P}$ denotes the total linear momentum of the mechanical system. On the other hand, when the Lagrangian is invariant under spatial rotations, $\mathbf{q} \to \mathbf{q} + (\delta \boldsymbol{\varphi} \times \mathbf{q})$ (where $\delta \boldsymbol{\varphi} = \delta \varphi \, \hat{\varphi}$ denotes a constant infinitesimal rotation about an axis along the $\hat{\varphi}$-direction), the Noether equation (2.44) yields the *angular* momentum conservation law

$$0 = \frac{d}{dt} \left(\mathbf{q} \times \frac{\partial L}{\partial \dot{\mathbf{q}}} \right) = \frac{d\mathbf{L}}{dt},$$

where $\mathbf{L} = \mathbf{q} \times \mathbf{P}$ denotes the total angular momentum of the mechanical system.

2.5.3 *Invariance Properties of a Lagrangian*

Lastly, an important invariance property of the Lagrangian is related to the fact that the Euler-Lagrange equations themselves are invariant under the *gauge* transformation

$$L \to L + \frac{dF}{dt} \tag{2.45}$$

on the Lagrangian itself, where $F(\mathbf{q}, t)$ is an arbitrary time-dependent function so that

$$\frac{dF(\mathbf{q}, t)}{dt} = \frac{\partial F}{\partial t} + \sum_j \dot{q}^j \frac{\partial F}{\partial q^j}.$$

To investigate the invariance property (2.45), we call $L' = L + dF/dt$ the new Lagrangian and L the old Lagrangian, and consider the new Euler-Lagrange equations

$$\frac{d}{dt}\left(\frac{\partial L'}{\partial \dot{q}^i}\right) = \frac{\partial L'}{\partial q^i}.$$

We now express each term in terms of the old Lagrangian L and the function F. Let us begin with

$$\frac{\partial L'}{\partial \dot{q}^i} = \frac{\partial}{\partial \dot{q}^i}\left(L + \frac{\partial F}{\partial t} + \sum_j \dot{q}^j \frac{\partial F}{\partial q^j}\right) = \frac{\partial L}{\partial \dot{q}^i} + \frac{\partial F}{\partial q^i},$$

so that

$$\frac{d}{dt}\left(\frac{\partial L'}{\partial \dot{q}^i}\right) = \frac{d}{dt}\left(\frac{\partial L}{\partial \dot{q}^i}\right) + \frac{\partial^2 F}{\partial t \partial q^i} + \sum_k \dot{q}^k \frac{\partial^2 F}{\partial q^k \partial q^i}.$$

Next, we find

$$\frac{\partial L'}{\partial q^i} = \frac{\partial}{\partial q^i}\left(L + \frac{\partial F}{\partial t} + \sum_j \dot{q}^j \frac{\partial F}{\partial q^j}\right)$$

$$= \frac{\partial L}{\partial q^i} + \frac{\partial^2 F}{\partial q^i \partial t} + \sum_j \dot{q}^j \frac{\partial^2 F}{\partial q^i \partial q^j}.$$

Using the symmetry properties

$$\dot{q}^j \frac{\partial^2 F}{\partial q^i \partial q^j} = \dot{q}^j \frac{\partial^2 F}{\partial q^j \partial q^i} \quad \text{and} \quad \frac{\partial^2 F}{\partial t \partial q^i} = \frac{\partial^2 F}{\partial q^i \partial t},$$

we easily verify that

$$\frac{d}{dt}\left(\frac{\partial L'}{\partial \dot{q}^i}\right) - \frac{\partial L'}{\partial q^i} = \frac{d}{dt}\left(\frac{\partial L}{\partial \dot{q}^i}\right) - \frac{\partial L}{\partial q^i} = 0.$$

Hence, since L and $L' = L + dF/dt$ lead to the same Euler-Lagrange equations, they are said to be equivalent.

Using this invariance property, for example, we note that the Lagrangian is also invariant under the Galilean velocity transformation $\mathbf{v} \to \mathbf{v} + \boldsymbol{\alpha}$, so that the Lagrangian variation

$$\delta L = \boldsymbol{\alpha} \cdot \left(\mathbf{v} \frac{\partial L}{\partial v^2}\right) \equiv \boldsymbol{\alpha} \cdot \frac{d\mathbf{x}}{dt} \frac{\partial L}{\partial v^2},$$

using the kinetic identity $\partial L/\partial v^2 = m/2$, can be written as an exact time derivative

$$\delta L = \frac{d}{dt}\left(\alpha \cdot \frac{m}{2}\mathbf{x}\right) \equiv \frac{d\delta F}{dt}.$$

Hence, because Lagrangian mechanics is invariant under the *gauge* transformation (2.45), the Lagrangian L is said to be Galilean invariant.

2.5.4 *Lagrangian Mechanics with Symmetries*

As an example of Lagrangian mechanics with symmetries, we return to the motion of a particle of mass m constrained to move on the surface of a cone of apex angle α (such that $\sqrt{x^2 + y^2} = z \tan \alpha$) in the presence of a gravitational field (see Fig. 2.4 and Sec. 2.3.3).

The Lagrangian for this constrained mechanical system is expressed in terms of the generalized coordinates (s, θ), where s denotes the distance from the cone's apex (labeled O in Fig. 2.4) and θ is the standard polar angle in the (x, y)-plane. Hence, by combining the kinetic energy $K = \frac{1}{2}m(\dot{s}^2 + s^2 \dot{\theta}^2 \sin^2 \alpha)$ with the potential energy $U = mgz = mg s \cos \alpha$, we construct the Lagrangian

$$L(s, \theta; \dot{s}, \dot{\theta}) = \frac{1}{2} m \left(\dot{s}^2 + s^2 \dot{\theta}^2 \sin^2 \alpha\right) - mg s \cos \alpha. \qquad (2.46)$$

Since the Lagrangian is independent of the polar angle θ, the canonical angular momentum

$$p_\theta = \frac{\partial L}{\partial \dot{\theta}} = ms^2 \dot{\theta} \sin^2 \alpha \qquad (2.47)$$

is a constant of the motion (as predicted by Noether's Theorem). The Euler-Lagrange equation for s, on the other hand, is expressed as

$$\ddot{s} + g \cos \alpha = s \dot{\theta}^2 \sin^2 \alpha = \frac{p_\theta^2}{m^2 s^3 \sin^2 \alpha}, \qquad (2.48)$$

where $g \cos \alpha$ denotes the component of the gravitational acceleration parallel to the surface of the cone. The right side of Eq. (2.48) involves s only after using $\dot{\theta} = p_\theta/(m s^2 \sin^2 \alpha)$, which follows from the conservation of angular momentum.

2.5.5 *Routh's Procedure*

Edward John Routh (1831-1907) introduced a simple procedure for eliminating ignorable degrees of freedom while introducing their corresponding conserved momenta within the context of Lagrangian Mechanics.

Consider, for example, two-dimensional motion on the (x, y)-plane represented by the Lagrangian $L(r; \dot{r}, \dot{\theta})$, where r and θ are the polar coordinates. Since the Lagrangian under consideration is independent of the angle θ, the canonical momentum $p_\theta = \partial L/\partial \dot{\theta}$ is conserved. Routh's procedure involves the construction of the *Routh-Lagrange* function (or Routhian)

$$R(r, \dot{r}; p_\theta) \equiv L(r; \dot{r}, \dot{\theta}) - p_\theta\, \dot{\theta}(r, p_\theta), \qquad (2.49)$$

where $\dot{\theta}$ is expressed as a function of r and p_θ. Note that the sign convention used in Eq. (2.49), which is different from Landau's convention [12], implies that the Routhian R is a *reduced* Lagrangian.

For example, for the constrained motion of a particle on the surface of a cone in the presence of gravity, the Lagrangian (2.46) can be reduced to the Routhian

$$R(s, \dot{s}; p_\theta) = \frac{1}{2}\, m\dot{s}^2 - \left(mg\, s\, \cos\alpha + \frac{p_\theta^2}{2ms^2 \sin^2\alpha}\right) = \frac{1}{2}\, m\dot{s}^2 - V(s),$$
$$(2.50)$$

and the equation of motion (2.48) can be expressed in Euler-Lagrange form

$$\frac{d}{ds}\left(\frac{\partial R}{\partial \dot{s}}\right) = \frac{\partial R}{\partial s} \quad \rightarrow \quad m\,\ddot{s} = -V'(s),$$

in terms of the effective potential

$$V(s) = mg\, s\, \cos\alpha + \frac{p_\theta^2}{2ms^2 \sin^2\alpha}.$$

Here, the effective potential $V(s)$ has a single minimum at $s = s_0$, where

$$s_0 = \left(\frac{p_\theta^2}{m^2 g\, \sin^2\alpha\, \cos\alpha}\right)^{\frac{1}{3}}$$

and $V_0 \equiv V(s_0) = \frac{3}{2}\, mg\, s_0 \cos\alpha$.

Figure 2.9 shows the results of the numerical integration of the dimensionless Euler-Lagrange equations for $\theta(\tau)$ and $\sigma(\tau) \equiv s(\tau)/s_0$, where $\tau \equiv t\,\sqrt{(g/s_0)}\, \cos\alpha$; see Appendix A.3 for some advice concerning the numerical solution of coupled ordinary differential equations. The top figure in Fig. 2.9 shows a projection of the path of the particle on the (x, z)-plane (side view), which clearly shows that the motion is periodic as the σ-coordinate oscillates between two finite values of σ. The bottom figure in Fig. 2.9 shows a projection of the path of the particle on the (x, y)-plane (top view), which shows the slow precession motion in the θ-coordinate.

In the next Chapter, we will show that the doubly-periodic motion of the particle moving on the surface of the inverted cone is a result of the conservation law of angular momentum and energy (since the Lagrangian system is also independent of time).

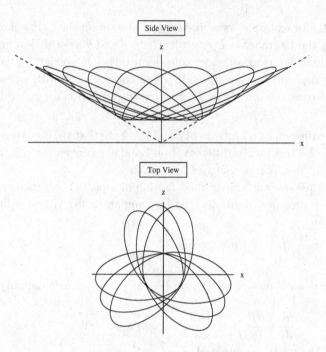

Fig. 2.9   Particle orbits on the surface of a cone.

2.6   Lagrangian Mechanics in the CM Frame

An important frame of reference associated with the dynamical description of the motion of interacting particles and rigid bodies is provided by the center-of-mass (CM) frame. The following discussion focuses on the Lagrangian for an isolated two-particle system expressed as

$$L = \frac{m_1}{2} |\dot{\mathbf{r}}_1|^2 + \frac{m_2}{2} |\dot{\mathbf{r}}_2|^2 - U(\mathbf{r}_1 - \mathbf{r}_2),$$

where $\mathbf{r}_1$ and $\mathbf{r}_2$ represent the positions of the particles of mass m_1 and m_2, respectively, and $U(\mathbf{r}_1, \mathbf{r}_2) = U(\mathbf{r}_1 - \mathbf{r}_2)$ is the potential energy for an isolated two-particle system (see Fig. 2.10).

Let us now define the position $\mathbf{R}$ of the center of mass

$$\mathbf{R} = \frac{m_1\,\mathbf{r}_1 + m_2\,\mathbf{r}_2}{m_1 + m_2},$$

and define the relative inter-particle position vector $\mathbf{r} = \mathbf{r}_1 - \mathbf{r}_2$, so that the particle positions can be expressed as

$$\mathbf{r}_1 = \mathbf{R} + \frac{m_2}{M}\mathbf{r} \quad \text{and} \quad \mathbf{r}_2 = \mathbf{R} - \frac{m_1}{M}\mathbf{r},$$

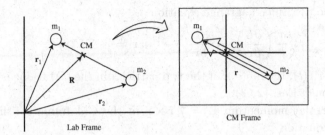

Fig. 2.10   Center-of-Mass frame.

where $M = m_1 + m_2$ is the total mass of the two-particle system (see Fig. 2.10). The Lagrangian of the isolated two-particle system, thus, becomes

$$L = \frac{M}{2}|\dot{\mathbf{R}}|^2 + \frac{\mu}{2}|\dot{\mathbf{r}}|^2 - U(\mathbf{r}),$$

where

$$\mu = \frac{m_1 m_2}{m_1 + m_2} = \left(\frac{1}{m_1} + \frac{1}{m_2}\right)^{-1}$$

denotes the *reduced* mass of the two-particle system. We note that the angular momentum of the two-particle system is expressed as

$$\mathbf{L} = \sum_a \mathbf{r}_a \times \mathbf{p}_a = \mathbf{R} \times \mathbf{P} + \mathbf{r} \times \mathbf{p}, \tag{2.51}$$

where the canonical momentum of the center-of-mass $\mathbf{P}$ and the canonical momentum $\mathbf{p}$ of the two-particle system in the CM frame are defined, respectively, as

$$\mathbf{P} = \frac{\partial L}{\partial \dot{\mathbf{R}}} = M\dot{\mathbf{R}} \quad \text{and} \quad \mathbf{p} = \frac{\partial L}{\partial \dot{\mathbf{r}}} = \mu\dot{\mathbf{r}}.$$

For an isolated system $(\partial L/\partial \mathbf{R} = 0)$, the canonical momentum $\mathbf{P}$ of the center-of-mass is a constant of the motion. The CM reference frame is defined by the condition $\mathbf{R} = 0$, i.e., we move the origin of our coordinate system to the CM position.

In the CM frame, the Lagrangian for an isolated two-particle system is

$$L(\mathbf{r}, \dot{\mathbf{r}}) = \frac{\mu}{2}|\dot{\mathbf{r}}|^2 - U(\mathbf{r}), \tag{2.52}$$

which describes the motion of a *fictitious* particle of mass μ at position $\mathbf{r}$, where the positions of the two *real* particles of masses m_1 and m_2 are

$$\mathbf{r}_1 = \frac{m_2}{M}\mathbf{r} \quad \text{and} \quad \mathbf{r}_2 = -\frac{m_1}{M}\mathbf{r}. \tag{2.53}$$

Hence, once the Euler-Lagrange equation

$$\frac{d}{dt}\left(\frac{\partial L}{\partial \dot{\mathbf{r}}}\right) = \frac{\partial L}{\partial \mathbf{r}} \quad \rightarrow \quad \mu \ddot{\mathbf{r}} = -\nabla U(\mathbf{r})$$

is solved for $\mathbf{r}(t)$, the motion of the two particles in the CM frame is determined through Eqs. (2.53).

The angular momentum $\mathbf{L} = \mu \mathbf{r} \times \dot{\mathbf{r}}$ in the CM frame satisfies the evolution equation

$$\frac{d\mathbf{L}}{dt} = \mathbf{r} \times \mu \ddot{\mathbf{r}} = -\mathbf{r} \times \nabla U(\mathbf{r}). \tag{2.54}$$

Here, using spherical coordinates (r, θ, φ), we find

$$\frac{d\mathbf{L}}{dt} = -\widehat{\varphi}\,\frac{\partial U}{\partial \theta} + \frac{\widehat{\theta}}{\sin\theta}\,\frac{\partial U}{\partial \varphi}.$$

If motion is originally taking place on the (x, y)-plane (i.e., at $\theta = \pi/2$) and the potential $U(r, \varphi)$ is independent of the polar angle θ, then the angular momentum vector is $\mathbf{L} = \ell\widehat{\mathbf{z}}$ and its magnitude ℓ satisfies the evolution equation

$$\frac{d\ell}{dt} = -\frac{\partial U}{\partial \varphi}.$$

Hence, for motion in a potential $U(r)$ that depends only on the radial position r, the angular momentum remains along the z-axis, and $\mathbf{L} = \ell\,\widehat{\mathbf{z}}$ represents an additional constant of motion. Motion in such central-force potentials will be studied in Chap. 4.

2.7 Problems

1. Consider a physical system composed of two blocks of mass m_1 and m_2 resting on incline planes placed at angles θ_1 and θ_2, respectively, as measured from the horizontal (see Fig. 2.11). The only active force acting on the blocks is due to gravity ($\mathbf{g} = -g\widehat{\mathbf{y}}$): $\mathbf{F}_i = -m_i g\widehat{\mathbf{y}}$ and, thus, the Principle of Virtual Work (2.13) implies that the system is in static equilibrium if

$$0 = m_1 g\widehat{\mathbf{y}} \cdot \delta\mathbf{x}^1 + m_2 g\widehat{\mathbf{y}} \cdot \delta\mathbf{x}^2.$$

Find the virtual displacements $\delta\mathbf{x}^1$ and $\delta\mathbf{x}^2$ needed to show that, according to the Principle of Virtual Work, the condition for static equilibrium is $m_1 \sin\theta_1 = m_2 \sin\theta_2$.

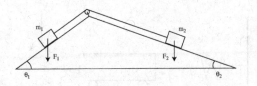

Fig. 2.11   Problem 1.

2. A particle of mass m is constrained to slide down a curve $y = V(x)$ under the action of gravity without friction. Show that the Euler-Lagrange equation for this system yields the equation

$$\ddot{x} = -V'\left(g + \ddot{V}\right),$$

where $\dot{V} = \dot{x}\,V'$ and $\ddot{V} = \ddot{x}\,V' + \dot{x}^2\,V''$.

3. Derive Eq. (2.40) for the compound Atwood machine.

4. A bead (of mass m) slides without friction on a wire in the shape of a cycloid: $x(\theta) = a\,(\theta - \sin\theta)$ and $y(\theta) = a\,(1 + \cos\theta)$.

(a) Find the Lagrangian $L(\theta, \dot{\theta})$ and derive the Euler-Lagrange equation for the angle θ.

(b) Find the equation of motion for $u = \cos(\theta/2)$.

5. A cart of mass M is placed on rails and attached to a wall with the help of a massless spring with constant k (Fig. 2.12); the spring is in its equilibrium state when the cart is at a distance x_0 from the wall. A pendulum of mass m and length ℓ is attached to the cart (as shown).

(a) Write the Lagrangian $L(x, \dot{x}, \theta, \dot{\theta})$ for the cart-pendulum system, where x denotes the position of the cart (as measured from a suitable origin) and θ denotes the angular position of the pendulum.

(b) From your Lagrangian, write the Euler-Lagrange equations for the generalized coordinates x and θ.

(c) Write the normalized equations for $\xi \equiv x/\ell$ and θ in terms of the normalized time $\tau \equiv t\sqrt{g/\ell}$ and the two dimensionless parameters $\mu \equiv m/(m + M) < 1$ and $\Omega^2 \equiv k\ell/[(m + M)g]$.

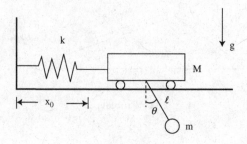

Fig. 2.12   Problem 5.

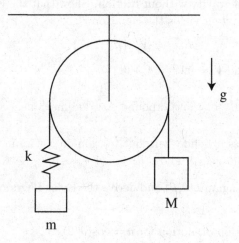

Fig. 2.13   Problem 6.

6. An Atwood machine is composed of two masses m and M attached by means of a massless rope into which a massless spring (with constant k) is inserted (as shown in Fig. 2.13). When the spring is in a relaxed state, the spring-rope length is ℓ.

(a) Find suitable generalized coordinates to describe the motion of the two masses (allowing for elongation or compression of the spring).

(b) Using these generalized coordinates, construct the Lagrangian and derive the appropriate Euler-Lagrange equations.

7. An Atwood machine is composed of two masses m and M attached by means of a massless rope. The massless pulley is attached to a massless spring with constant k (as shown in Fig. 2.14).

(a) Find suitable generalized coordinates to describe the motion of the two masses (allowing for elongation or compression of the spring).

(b) Using these generalized coordinates, construct the Lagrangian and derive the appropriate Euler-Lagrange equations.

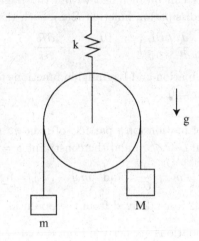

Fig. 2.14   Problem 7.

8. A particle of mass m slides down on a smooth circular wedge of mass M. The wedge rests on a smooth horizontal table.

(a) Find the equations of motion for m and M.

(b) Find the reaction of the wedge on m.

9. A pendulum of length ℓ and mass m is attached to a point of mass M that is constrained to only move horizontally.

(a) Derive the coupled Euler-Lagrange equations for the horizontal displacement x and the angular displacement θ.

(b) Show that this system possesses a symmetry related to translations along the x-axis. Using the corresponding conservation law, show that the coupled equations derived in Part (a) can be expressed as

$$\left(1 - \mu \, \cos^2 \theta\right) \, \ddot{\theta} + \omega_g^2 \, \sin \theta + \frac{\mu}{2} \, \dot{\theta}^2 \, \sin 2\theta = 0,$$

where $\mu \equiv m/M$ is the mass ratio and $\omega_g \equiv \sqrt{g/\ell}$ is the pendulum angular frequency.

10. Dissipative effects can be included within the Lagrangian formalism through the Rayleigh dissipation function $\mathcal{R}(\dot{x})$, such that

$$\frac{d}{dt}\left(\frac{\partial L}{\partial \dot{x}}\right) - \frac{\partial L}{\partial x} = -\frac{\partial \mathcal{R}}{\partial \dot{x}}.$$

Find the Rayleigh dissipation and Lagrangian functions for the equation of motion $m\ddot{x} + \lambda \dot{x} + kx = 0$.

11*. The equations of motion for a particle of mass m moving under the influence of a potential $U(x, y, z)$ and the constraint $z = f(x, y)$ are

$$m\ddot{x} - F_x = (F_z - m\ddot{z})\frac{\partial f}{\partial x} \quad \text{and} \quad m\ddot{y} - F_y = (F_z - m\ddot{z})\frac{\partial f}{\partial y},$$

where $\mathbf{F} \equiv -\nabla U$ is the force derived from the potential U.

(a) Show that these equations are derived from the constrained Lagrangian $L = \frac{1}{2}m|\dot{\mathbf{x}}|^2 - U - \lambda\,[z - f(x, y)]$.

(b) Show that these equations follow from Gauss' Least Constraint Principle

$$\delta\,|m\,\ddot{\mathbf{x}} - \mathbf{F}|^2 \equiv (m\ddot{\mathbf{x}} - \mathbf{F}) \cdot m\,\delta\ddot{\mathbf{x}} = 0,$$

where the variation is applied with respect to the acceleration $\ddot{\mathbf{x}}$ only.

12*. Rocket propulsion is described in terms of the equations of motion $\ddot{\mathbf{x}} + \nabla\Phi = (\dot{m}/m)\,\mathbf{c}$, where $\mathbf{c}$ is the exhaust velocity relative to the rocket and $\Phi(\mathbf{x})$ is the potential energy per unit mass. By definition, the mass loss rate is given as $\dot{m}/m = -\frac{1}{c}|\ddot{\mathbf{x}} + \nabla\Phi| \equiv -a/c$, which implies that, if the magnitude $c \equiv |\mathbf{c}|$ of the exhaust velocity is constant, the ratio m_f/m_i of the final mass to the initial mass of the rocket is expressed as

$$\frac{m_f}{m_i} = \exp\left(-\frac{1}{c}\int_{t_i}^{t_f} a(t)\,dt\right).$$

The mass ratio is therefore maximum (i.e., the rocket uses the least amount of fuel) if the integral

$$\int_{t_i}^{t_f} a(\mathbf{x}, \ddot{\mathbf{x}})\, dt \equiv \int_{t_i}^{t_f} |\ddot{\mathbf{x}} + \nabla\Phi(\mathbf{x})|\, dt$$

is minimum.

(a) Show that the Euler-Lagrange equation for this problem is

$$\frac{d^2}{dt^2}\left(\frac{\partial a}{\partial \ddot{\mathbf{x}}}\right) + \frac{\partial a}{\partial \mathbf{x}} = 0,$$

and derive that equation in terms of the unit vector $\hat{c} \equiv \mathbf{c}/c$.

(b) By using the fact that $\hat{c}$ is a unit vector, show that the Euler-Lagrange equation derived in Part (a) can be written as $|d\hat{c}/dt|^2 = \hat{c} \cdot \nabla\nabla\Phi \cdot \hat{c}$, so that a minimum solution exists provided the condition $\hat{c} \cdot \nabla\nabla\Phi \cdot \hat{c} \geq 0$ is satisfied (where the equality applies to the case $d\hat{c}/dt \equiv 0$).

(c) Show that the minimum condition $\hat{c} \cdot \nabla\nabla\Phi \cdot \hat{c} \geq 0$ yields $\cos\psi \equiv \hat{c} \cdot \hat{r} \leq 1/\sqrt{3}$ for the case of the attractive gravitational potential $\Phi(\mathbf{x}) = -GM/r$, where M denotes the mass of the object to which the rocket is attracted.

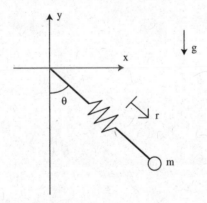

Fig. 2.15   Problem 13.

13. An oscillating pendulum consists of a bob of mass m attached to a spring of constant k and relaxed length l_0 (see Fig. 2.15). The generalized coordinates for this system are the angle θ and the displacement r away from the spring's equilibrium. Find the Lagrangian $L(r, \theta; \dot{r}, \dot{\theta})$ and derive the Euler-Lagrange equations for r and θ.

Chapter 3

Hamiltonian Mechanics

In the previous Chapter, the Lagrangian method was introduced as a powerful alternative to the Newtonian method for deriving equations of motion for multi-particle mechanical systems. In the present Chapter, a complementary approach to the Lagrangian method, known as the Hamiltonian method, is presented. Although much of the Hamiltonian method is outside the scope of this course (e.g., the canonical and noncanonical Hamiltonian formulations of Classical Mechanics and the Hamiltonian formulation of Quantum Mechanics), a simplified version (the Energy method) is presented here as a practical method for *solving* the Euler-Lagrange equations by quadrature. See Appendix C for a brief introduction to the modern formulation of Hamiltonian Mechanics.

3.1 Canonical Hamilton's Equations

The k second-order Euler-Lagrange equations on configuration space $\mathbf{q}$:

$$\frac{d}{dt}\left(\frac{\partial L}{\partial \dot{q}^j}\right) = \frac{\partial L}{\partial q^j}, \tag{3.1}$$

can be written as $2k$ first-order differential equations on a $2k$-dimensional *phase* space with coordinates $\mathbf{z} = (q^1, ..., q^k;\ p_1, ..., p_k)$, where

$$p_j(\mathbf{q}, \dot{\mathbf{q}}; t) = \frac{\partial L}{\partial \dot{q}^j}(\mathbf{q}, \dot{\mathbf{q}}; t) \tag{3.2}$$

defines the jth-component of the *canonical* momentum. In terms of these new coordinates, the Euler-Lagrange equations (3.1) are transformed into Hamilton's canonical equations (William Rowan Hamilton, 1805-1865)

$$\frac{dq^j}{dt} = \frac{\partial H}{\partial p_j} \quad \text{and} \quad \frac{dp_j}{dt} = -\frac{\partial H}{\partial q^j}, \tag{3.3}$$

73

where the Hamiltonian function $H(\mathbf{q}, \mathbf{p}; t)$ is defined from the Lagrangian function $L(\mathbf{q}, \dot{\mathbf{q}}; t)$ by the Legendre transformation (Adrien-Marie Legendre, 1752-1833)

$$H(\mathbf{q}, \mathbf{p}; t) = \mathbf{p} \cdot \dot{\mathbf{q}}(\mathbf{q}, \mathbf{p}, t) - L[\mathbf{q}, \dot{\mathbf{q}}(\mathbf{q}, \mathbf{p}, t), t]. \tag{3.4}$$

We note that Hamilton's *canonical* equations of motion (3.3) are completely equivalent to the Lagrangian formulation.

We note that Hamilton's equations (3.3) can also be derived from a variational principle in phase space $\mathbf{z} = (\mathbf{q}, \mathbf{p})$ as follows. First, we use the inverse of the Legendre transformation

$$L(\mathbf{z}, \dot{\mathbf{z}}; t) = \mathbf{p} \cdot \dot{\mathbf{q}} - H(\mathbf{z}; t) \tag{3.5}$$

to obtain an expression for the Lagrangian function in phase space. Next, we calculate the first-variation of the action integral

$$\delta \int L(\mathbf{q}, \mathbf{p}; t) \, dt = \int \left[\delta\mathbf{p} \cdot \left(\dot{\mathbf{q}} - \frac{\partial H}{\partial \mathbf{p}} \right) + \left(\mathbf{p} \cdot \delta\dot{\mathbf{q}} - \delta\mathbf{q} \cdot \frac{\partial H}{\partial \mathbf{q}} \right) \right] dt,$$

where the variations δq^i and δp_i are now considered independent (and they are both assumed to vanish at the end points). By integrating by parts the term $\mathbf{p} \cdot \delta\dot{\mathbf{q}}$, we find

$$\delta \int L(\mathbf{q}, \mathbf{p}; t) \, dt = \int \left[\delta\mathbf{p} \cdot \left(\dot{\mathbf{q}} - \frac{\partial H}{\partial \mathbf{p}} \right) - \delta\mathbf{q} \cdot \left(\dot{\mathbf{p}} + \frac{\partial H}{\partial \mathbf{q}} \right) \right] dt,$$

so that the Principle of Least Action $\int \delta L \, dt = 0$ now yields Hamilton's equations (3.3) for arbitrary variations $(\delta\mathbf{q}, \delta\mathbf{p})$.

Lastly, an important equation associated with Hamilton's principal function $\mathcal{S}$ can be derived from the infinitesimal action

$$d\mathcal{S}(\mathbf{q}, t) = \mathbf{p} \cdot d\mathbf{q} - H \, dt, \tag{3.6}$$

from which we obtain the relations

$$\left. \begin{array}{c} H = -\partial \mathcal{S}/\partial t \\[2mm] \mathbf{p} = \partial \mathcal{S}/\partial \mathbf{q} \end{array} \right\}. \tag{3.7}$$

These relations can be used to obtain the Hamilton-Jacobi equation for particle dynamics [7]

$$\frac{\partial \mathcal{S}}{\partial t} + H\left(\mathbf{q}, \frac{\partial \mathcal{S}}{\partial \mathbf{q}}; t \right) = 0. \tag{3.8}$$

The solution to this equation is said to generate a canonical transformation that annihilates the Hamiltonian, i.e., the function $\mathcal{S}$ generates a time-dependent canonical transformation $\mathbf{z} = (\mathbf{q}, \mathbf{p}) \rightarrow \mathbf{Z} = (\mathbf{Q}, \mathbf{P})$ such that the new Hamiltonian $K(\mathbf{Z}; t) \equiv H(\mathbf{z}(\mathbf{Z}, t); t) + \partial_t S(\mathbf{q}(\mathbf{Z}, t), t)$ vanishes. Applications of the Hamilton-Jacobi equation fall outside the scope of the present course [7, 12]. We simply mention here that the Hamilton-Jacobi equation (3.8) figures prominently in the historical connection between particle dynamics and wave mechanics (as discussed in Sec. 3.3), as well as the connection between Classical Mechanics and Quantum Mechanics (as discussed in Sec. 9.3 and problem 1 in Chap. 9).

3.2 Legendre Transformation*

Before proceeding with the Hamiltonian formulation of particle dynamics, we investigate the condition under which the Legendre transformation (3.4) is possible. It turns out that this condition is associated with the condition under which the inversion of the relation $\mathbf{p}(\mathbf{r}, \dot{\mathbf{r}}, t) \rightarrow \dot{\mathbf{r}}(\mathbf{r}, \mathbf{p}, t)$ is possible. To simplify our discussion, we focus on motion in two dimensions (labeled x and y).

The general expression of the kinetic energy term of a Lagrangian with two degrees of freedom $L(x, \dot{x}, y, \dot{y}) = K(x, \dot{x}, y, \dot{y}) - U(x, y)$ is

$$K(x, \dot{x}, y, \dot{y}) = \frac{\alpha}{2} \dot{x}^2 + \beta \dot{x}\dot{y} + \frac{\gamma}{2} \dot{y}^2 = \frac{1}{2} \dot{\mathbf{r}}^\mathsf{T} \cdot \mathsf{M} \cdot \dot{\mathbf{r}}, \qquad (3.9)$$

where $\dot{\mathbf{r}}^\mathsf{T} = (\dot{x}, \dot{y})$ denotes the transpose of $\dot{\mathbf{r}}$ (see Appendix A for additional details concerning linear algebra) and the *mass* matrix M is

$$\mathsf{M} = \begin{pmatrix} \alpha & \beta \\ \beta & \gamma \end{pmatrix}.$$

Here, the coefficients α, β, and γ may be functions of x and y. The canonical momentum vector (3.2) is thus defined as

$$\mathbf{p} = \frac{\partial L}{\partial \dot{\mathbf{r}}} = \mathsf{M} \cdot \dot{\mathbf{r}} \rightarrow \begin{pmatrix} p_x \\ p_y \end{pmatrix} = \begin{pmatrix} \alpha & \beta \\ \beta & \gamma \end{pmatrix} \cdot \begin{pmatrix} \dot{x} \\ \dot{y} \end{pmatrix}$$

or

$$\left. \begin{aligned} p_x &= \alpha \dot{x} + \beta \dot{y} \\ p_y &= \beta \dot{x} + \gamma \dot{y} \end{aligned} \right\}. \qquad (3.10)$$

The Lagrangian is said to be *regular* if the mass matrix M is invertible, i.e., if its determinant

$$\Delta = \alpha\gamma - \beta^2 \neq 0.$$

In the case of a regular Lagrangian, we readily invert (3.10) to obtain

$$\dot{\mathbf{r}}(\mathbf{r}, \mathbf{p}, t) = \mathsf{M}^{-1} \cdot \mathbf{p} \rightarrow \begin{pmatrix} \dot{x} \\ \dot{y} \end{pmatrix} = \frac{1}{\Delta} \begin{pmatrix} \gamma & -\beta \\ -\beta & \alpha \end{pmatrix} \cdot \begin{pmatrix} p_x \\ p_y \end{pmatrix}$$

or

$$\left. \begin{aligned} \dot{x} &= (\gamma\, p_x - \beta\, p_y)/\Delta \\ \dot{y} &= (\alpha\, p_y - \beta\, p_x)/\Delta \end{aligned} \right\}, \tag{3.11}$$

and the kinetic energy term becomes

$$K(x, p_x,\, y, p_y) = \frac{1}{2}\, \mathbf{p}^{\mathsf{T}} \cdot \mathsf{M}^{-1} \cdot \mathbf{p}.$$

Lastly, under the Legendre transformation (3.4), we find

$$H = \mathbf{p}^{\mathsf{T}} \cdot (\mathsf{M}^{-1} \cdot \mathbf{p}) - \left(\frac{1}{2}\, \mathbf{p}^{\mathsf{T}} \cdot \mathsf{M}^{-1} \cdot \mathbf{p} - U \right)$$

$$= \frac{1}{2}\, \mathbf{p}^{\mathsf{T}} \cdot \mathsf{M}^{-1} \cdot \mathbf{p} + U.$$

Hence, we clearly see that the Legendre transformation is applicable only if the mass matrix M in the kinetic energy (3.9) is invertible. Lastly, we note that the Legendre transformation is also used in other areas in physics such as Thermodynamics.

3.3   Hamiltonian Optics and Wave-Particle Duality*

Historically, the Hamiltonian method was first introduced as a formulation of the dynamics of light rays [4, 18]. Consider the following *phase* integral

$$\Theta[\mathbf{z}] \equiv \int_{t_1}^{t_2} \theta(\mathbf{x}, \mathbf{k}; t)\, dt = \int_{t_1}^{t_2} [\, \mathbf{k} \cdot \dot{\mathbf{x}} - \omega(\mathbf{x}, \mathbf{k}; t) \,]\, dt, \tag{3.12}$$

where $\Theta[\mathbf{z}]$ is a functional of the light-path $\mathbf{z}(t) = (\mathbf{x}(t), \mathbf{k}(t))$ in *ray phase space*, expressed in terms of the instantaneous position $\mathbf{x}(t)$ of a light ray and its associated instantaneous wave vector $\mathbf{k}(t)$; here, the dispersion relation $\omega(\mathbf{x}, \mathbf{k}; t)$ is obtained as a root of the dispersion equation $\det \mathsf{D}(\mathbf{x}, t; \mathbf{k}, \omega) = 0$, and a dot denotes a total time derivative: $\dot{\mathbf{x}} = d\mathbf{x}/dt$.

Assuming that the phase integral $\Theta[\mathbf{z}]$ acquires a *stationary* value for a *physical* ray orbit $\mathbf{z}(t)$, henceforth called the Principle of *Stationary Phase*

Table 3.1   Correspondence between Particle Mechanics and Wave Mechanics.

	Particle	Wave
Phase Space	$\mathbf{z} = (\mathbf{q}, \mathbf{p})$	$\mathbf{z} = (\mathbf{x}, \mathbf{k})$
Hamiltonian	$H(\mathbf{z}; t)$	$\omega(\mathbf{z}; t)$
Variational Principle I	Maupertuis-Jacobi	Fermat
$\int (\cdots) \, ds$	$\sqrt{2m(E - U)}$	n
Variational Principle II	Hamilton	Stationary Phase
$\int (\cdots) \, dt$	$L = \mathbf{p} \cdot \dot{\mathbf{q}} - H$	$\theta = \mathbf{k} \cdot \dot{\mathbf{x}} - \omega$
Hamilton-Jacobi	$\partial_t \mathcal{S} + H(\mathbf{q}, \partial_{\mathbf{q}} \mathcal{S}, t) = 0$	$\partial_t \Theta + \omega(\mathbf{x}, \nabla \Theta, t) = 0$
Hamilton's equations	$(\dot{\mathbf{q}}, \dot{\mathbf{p}}) = (\partial_{\mathbf{p}} H, -\partial_{\mathbf{q}} H)$	$(\dot{\mathbf{x}}, \dot{\mathbf{k}}) = (\partial_{\mathbf{k}} \omega, -\nabla \omega)$

$\delta\Theta = 0$, we can show that Euler's First Equation leads to Hamilton's (canonical) ray equations:

$$\frac{d\mathbf{x}}{dt} = \frac{\partial\omega}{\partial\mathbf{k}} \quad \text{and} \quad \frac{d\mathbf{k}}{dt} = -\nabla\omega. \tag{3.13}$$

The first ray equation states that a ray travels at the *group* velocity while the second ray equation states that the wave vector $\mathbf{k}$ is refracted as the ray propagates in a non-uniform medium (see Chap. 1). Hence, the frequency function $\omega(\mathbf{x}, \mathbf{k}; t)$ is the Hamiltonian of ray dynamics in a nonuniform medium.

The Hamilton-Jacobi equation for ray optics is obtained from the infinitesimal phase

$$d\Theta(\mathbf{x}, t) = \mathbf{k} \cdot d\mathbf{x} - \omega \, dt, \tag{3.14}$$

from which we obtain the *eikonal* relations

$$\left. \begin{array}{l} \omega = -\partial\Theta/\partial t \\[2mm] \mathbf{k} = \nabla\Theta \end{array} \right\}. \tag{3.15}$$

These relations can then be used to obtain the Hamilton-Jacobi equation

$$\frac{\partial\Theta}{\partial t} + \omega(\mathbf{x}, \nabla\Theta; t) = 0, \tag{3.16}$$

where $\omega(\mathbf{x}, \mathbf{k}; t)$ is the Hamiltonian for the ray equations (3.13). The analogy between the Hamilton-Jacobi equation (3.8) for particle dynamics and the Hamilton-Jacobi equation (3.16) for ray optics leads us to recognize the deep connections between Classical Mechanics and Wave Mechanics (see Table 3.1 for a detailed correspondence).

It was de Broglie who noted (as a graduate student well versed in Classical Mechanics) the similarities between Hamilton's equations (3.3) and (3.13), on the one hand, and the Maupertuis-Jacobi (2.1) and Euler-Lagrange (2.27) Principles of Least Action and Fermat's Principle of Least Time (1.35) and Principle of Stationary Phase (3.12), on the other hand (see Table 3.1). By using the *quantum of action* $\hbar = h/2\pi$ defined in terms of Planck's constant h and Planck's energy hypothesis $E = \hbar\omega$, de Broglie suggested that a particle's momentum $\mathbf{p}$ be related to its wavevector $\mathbf{k}$ according to de Broglie's formula $\mathbf{p} = \hbar\mathbf{k}$ and introduced the wave-particle synthesis based on the identity

$$S[\mathbf{z}] = \hbar\Theta[\mathbf{z}] \tag{3.17}$$

involving the action integral $S[\mathbf{z}]$ and the phase integral $\Theta[\mathbf{z}]$.

The final synthesis between Classical and Quantum Mechanics came from Richard Phillips Feynman (1918-1988) who provided an explicit derivation of Schroedinger's equation (Erwin Rudolf Josef Alexander Schroedinger, 1887-1961) by associating the probability that a particle follow a particular path $\mathbf{z}(t; \mathbf{z}_0)$ with the expression $\exp(i\,\hbar^{-1}S[\mathbf{z}])$, where $S[\mathbf{z}]$ denotes the action integral for the path [18].

3.4 Motion in an Electromagnetic Field

Although the problem of the motion of a charged particle in an electromagnetic field can be considered outside the scope of the present course, it represents a important paradigm that beautifully illustrates the connection between Lagrangian and Hamiltonian mechanics and it is well worth studying.

3.4.1 *Euler-Lagrange Equations*

The equations of motion for a charged particle of mass m and charge e moving in an electromagnetic field represented by the electric field $\mathbf{E}$ and magnetic field $\mathbf{B}$ are

$$\frac{d\mathbf{x}}{dt} = \mathbf{v} \tag{3.18}$$

$$\frac{d\mathbf{v}}{dt} = \frac{e}{m}\left(\mathbf{E} + \frac{d\mathbf{x}}{dt} \times \frac{\mathbf{B}}{c}\right), \tag{3.19}$$

where $\mathbf{x}$ denotes the position of the particle and $\mathbf{v}$ its velocity (Note: Gaussian units are used whenever electromagnetic fields are involved).

By treating the coordinates $(\mathbf{x}, \mathbf{v})$ as generalized coordinates (i.e., $\delta\mathbf{v}$ is treated independently from $\delta\mathbf{x}$), we can show that the equations of motion

(3.18) and (3.19) can be obtained as Euler-Lagrange equations from the Lagrangian (3.5):

$$L(\mathbf{x}, \dot{\mathbf{x}}, \mathbf{v}, \dot{\mathbf{v}}; t) = \left(m\mathbf{v} + \frac{e}{c}\mathbf{A}(\mathbf{x}, t) \right) \cdot \dot{\mathbf{x}} - \left(e\Phi(\mathbf{x}, t) + \frac{m}{2}|\mathbf{v}|^2 \right),$$

$$(3.20)$$

where Φ and $\mathbf{A}$ are the electromagnetic potentials in terms of which electric and magnetic fields are defined

$$\mathbf{E} = -\nabla\Phi - \frac{1}{c}\frac{\partial\mathbf{A}}{\partial t} \quad \text{and} \quad \mathbf{B} = \nabla \times \mathbf{A}. \tag{3.21}$$

Note that these expressions for $\mathbf{E}$ and $\mathbf{B}$ satisfy Faraday's law $\nabla \times \mathbf{E} = -c^{-1}\partial\mathbf{B}/\partial t$ and the divergenceless property $\nabla \cdot \mathbf{B} = 0$ of the magnetic field.

First, we look at the Euler-Lagrange equation for $\mathbf{x}$:

$$\frac{\partial L}{\partial\dot{\mathbf{x}}} = m\mathbf{v} + \frac{e}{c}\mathbf{A} \quad \rightarrow \quad \frac{d}{dt}\left(\frac{\partial L}{\partial\dot{\mathbf{x}}}\right) = m\dot{\mathbf{v}} + \frac{e}{c}\left(\frac{\partial\mathbf{A}}{\partial t} + \dot{\mathbf{x}}\cdot\nabla\mathbf{A}\right)$$

$$\frac{\partial L}{\partial\mathbf{x}} = \frac{e}{c}\nabla\mathbf{A}\cdot\dot{\mathbf{x}} - e\nabla\Phi,$$

which yields the Lorentz force equation (3.19), since

$$m\dot{\mathbf{v}} = -e\left(\nabla\Phi + \frac{1}{c}\frac{\partial\mathbf{A}}{\partial t}\right) + \frac{e}{c}\dot{\mathbf{x}}\times\nabla\times\mathbf{A} = e\mathbf{E} + \frac{e}{c}\dot{\mathbf{x}}\times\mathbf{B}, \tag{3.22}$$

where the definitions (3.21) were used.

Next, we look at the Euler-Lagrange equation for $\mathbf{v}$:

$$\frac{\partial L}{\partial\dot{\mathbf{v}}} = 0 \quad \rightarrow \quad \frac{d}{dt}\left(\frac{\partial L}{\partial\dot{\mathbf{v}}}\right) = 0 = \frac{\partial L}{\partial\mathbf{v}} = m\dot{\mathbf{x}} - m\mathbf{v},$$

which yields Eq. (3.18). Because $\partial L/\partial\dot{\mathbf{v}} = 0$, we note that we could use Eq. (3.18) as a constraint which could be imposed *a priori* on the Lagrangian (3.20) to give

$$L(\mathbf{x}, \dot{\mathbf{x}}; t) = \frac{m}{2}|\dot{\mathbf{x}}|^2 + \frac{e}{c}\mathbf{A}(\mathbf{x}, t)\cdot\dot{\mathbf{x}} - e\Phi(\mathbf{x}, t). \tag{3.23}$$

The Euler-Lagrange equation for $\mathbf{x}$ in this case is identical to Eq. (3.22) with $\dot{\mathbf{v}} = \ddot{\mathbf{x}}$.

3.4.2 *Energy Conservation Law*

We now show that the second Euler equation (i.e., the energy conservation law), expressed as

$$\frac{d}{dt}\left(L - \dot{\mathbf{x}}\cdot\frac{\partial L}{\partial\dot{\mathbf{x}}}\right) = \frac{\partial L}{\partial t},$$

is satisfied exactly by the Lagrangian (3.23) and the equations of motion (3.18) and (3.19). First, from the Lagrangian (3.23), we find

$$\frac{\partial L}{\partial t} = \frac{e}{c}\frac{\partial \mathbf{A}}{\partial t} \cdot \mathbf{v} - e\frac{\partial \Phi}{\partial t}$$

$$L - \dot{\mathbf{x}} \cdot \frac{\partial L}{\partial \dot{\mathbf{x}}} = L - \left(m\mathbf{v} + \frac{e}{c}\mathbf{A}\right)\cdot \mathbf{v}$$

$$= -\left(\frac{m}{2}|\mathbf{v}|^2 + e\Phi\right).$$

Next, we find

$$\frac{d}{dt}\left(L - \dot{\mathbf{x}}\cdot\frac{\partial L}{\partial \dot{\mathbf{x}}}\right) = -m\mathbf{v}\cdot\dot{\mathbf{v}} - e\left(\frac{\partial \Phi}{\partial t} + \mathbf{v}\cdot\nabla\Phi\right)$$

$$= -e\frac{\partial}{\partial t}\left(\Phi - \frac{\mathbf{v}}{c}\cdot\mathbf{A}\right).$$

Using Eq. (3.18), we readily find $m\mathbf{v}\cdot\dot{\mathbf{v}} = e\,\mathbf{E}\cdot\mathbf{v}$ and thus

$$-e\,\mathbf{E}\cdot\mathbf{v} - e\left(\frac{\partial \Phi}{\partial t} + \mathbf{v}\cdot\nabla\Phi\right) = \frac{e}{c}\frac{\partial \mathbf{A}}{\partial t}\cdot\mathbf{v} - e\frac{\partial \Phi}{\partial t},$$

which is shown to be satisfied exactly by substituting the definition (3.21) for $\mathbf{E}$.

3.4.3 *Gauge Invariance*

The electric and magnetic fields defined in (3.21) are invariant under the gauge transformation

$$\Phi \rightarrow \Phi - \frac{1}{c}\frac{\partial \chi}{\partial t} \quad \text{and} \quad \mathbf{A} \rightarrow \mathbf{A} + \nabla\chi, \tag{3.24}$$

where $\chi(\mathbf{x}, t)$ is an arbitrary scalar field. Although the equations of motion (3.18) and (3.19) are *manifestly* gauge invariant, the Lagrangian (3.23) is not manifestly gauge invariant since the electromagnetic potentials Φ and $\mathbf{A}$ appear explicitly. Under a gauge transformation (3.24), however, we find

$$L \rightarrow L + \frac{e}{c}\dot{\mathbf{x}}\cdot\nabla\chi - e\left(-\frac{1}{c}\frac{\partial \chi}{\partial t}\right) = L + \frac{d}{dt}\left(\frac{e}{c}\chi\right).$$

Since Lagrangian Mechanics is invariant under the transformation (2.45), the Lagrangian (3.23) is invariant under the gauge transformation (3.24).

3.4.4 *Canonical Hamilton's Equations*

The canonical momentum $\mathbf{p}$ for a particle of mass m and charge e in an electromagnetic field is defined as

$$\mathbf{p}(\mathbf{x}, \mathbf{v}, t) = \frac{\partial L}{\partial \dot{\mathbf{x}}} = m\,\mathbf{v} + \frac{e}{c}\,\mathbf{A}(\mathbf{x}, t), \qquad (3.25)$$

which is the sum of the kinetic momentum $(m\,\mathbf{v})$ and a magnetic contribution (represented by the vector potential $\mathbf{A}$). The canonical Hamiltonian function $H(\mathbf{x}, \mathbf{p}, t)$ is now constructed through the Legendre transformation

$$H(\mathbf{x}, \mathbf{p}, t) = \mathbf{p} \cdot \dot{\mathbf{x}}(\mathbf{x}, \mathbf{p}, t) - L[\mathbf{x}, \dot{\mathbf{x}}(\mathbf{x}, \mathbf{p}, t), t]$$

$$= e\,\Phi(\mathbf{x}, t) + \frac{1}{2m}\left| \mathbf{p} - \frac{e}{c}\,\mathbf{A}(\mathbf{x}, t) \right|^2, \qquad (3.26)$$

where $\mathbf{v}(\mathbf{x}, \mathbf{p}, t)$ was obtained by inverting $\mathbf{p}(\mathbf{x}, \mathbf{v}, t)$ from Eq. (3.25). Using the canonical Hamiltonian function (3.26), we immediately find

$$\dot{\mathbf{x}} = \frac{\partial H}{\partial \mathbf{p}} = \frac{1}{m}\left(\mathbf{p} - \frac{e}{c}\,\mathbf{A} \right),$$

$$\dot{\mathbf{p}} = -\frac{\partial H}{\partial \mathbf{x}} = -e\,\nabla\Phi - \frac{e}{c}\,\nabla\mathbf{A} \cdot \dot{\mathbf{x}},$$

from which we recover the equations of motion (3.18) and (3.19) once we use the definition (3.25) for the canonical momentum.

We should warn the reader that the simplicity of the canonical Hamiltonian formalism comes at a price: the canonical momentum $\mathbf{p} \equiv \partial L/\partial \dot{\mathbf{q}}$ and the Hamiltonian are not physical quantities. Indeed, under the gauge transformation (3.24), the canonical momentum and Hamiltonian are transformed as

$$\mathbf{p} \to \mathbf{p} + \frac{e}{c}\,\nabla\chi \quad \text{and} \quad H \to H - \frac{e}{c}\,\frac{\partial\chi}{\partial t}.$$

These transformations, however, leave Hamilton's canonical equations invariant.

3.4.5 *Maupertuis' Principle of Least Action*

Because of the close connection between Fermat's Principle of Least Time and Maupertuis' Principle of Least Action (see Table 3.1), it is instructive to derive a variational principle suitable for applications in electron *optics*, where the path of an electron beam is guided by electric and magnetic lenses.

We begin with the classical action integral

$$\int L \, dt = \int \left(L - \frac{\partial L}{\partial \dot{\mathbf{r}}} \cdot \dot{\mathbf{r}} \right) dt + \int \frac{\partial L}{\partial \dot{\mathbf{r}}} \cdot d\mathbf{r}$$

$$= - \int H \, dt + \int \mathbf{p} \cdot d\mathbf{r},$$

where we used the Legendre transformation (3.4) in the first integral and used the definition (3.2) for the canonical momentum in the second integral. If we consider time-independent guiding magnetic fields, the electron energy is constant $(H = E)$ and Maupertuis' Principle of Least Action $\delta S_E = 0$ is expressed in terms of the action functional (at constant energy E) [3]

$$S_E = \int \mathbf{p} \cdot d\mathbf{r} = \int \left(m\mathbf{v} + \frac{q}{c} \mathbf{A} \right) \cdot d\mathbf{r}$$

$$= \int \left(mv + \frac{q}{c} \mathbf{A} \cdot \widehat{\mathbf{s}} \right) ds, \tag{3.27}$$

where electrons have mass m and charge $q = -e$, the unit vector $\widehat{\mathbf{s}}$ is defined as $\widehat{\mathbf{s}} \equiv d\mathbf{r}/ds$, and $mv \equiv \sqrt{2m(E - q\Phi)}$ is defined in terms of the total energy E and the electric scalar potential Φ.

3.5　One-degree-of-freedom Hamiltonian Dynamics

In this Section, we investigate Hamiltonian dynamics with one degree of freedom in a time-independent potential. In particular, we show that such systems are always integrable (i.e., they can always be solved by quadrature).

3.5.1　*Energy Method*

The one degree-of-freedom Hamiltonian dynamics of a particle of mass m is based on the Hamiltonian

$$H(x, p) = \frac{p^2}{2m} + U(x), \tag{3.28}$$

where $p = m\dot{x}$ is the particle's momentum and $U(x)$ is the time-independent potential energy. The Hamilton's equations (3.3) for this Hamiltonian are

$$\frac{dx}{dt} = \frac{p}{m} \quad \text{and} \quad \frac{dp}{dt} = - \frac{dU(x)}{dx}. \tag{3.29}$$

Since the Hamiltonian (and Lagrangian) is time independent, the energy conservation law states that $H(x, p) = E$. In turn, this conservation law implies that the particle's velocity $\dot{x}$ can be expressed as

$$\dot{x}(x, E) = \pm\sqrt{\frac{2}{m}\left[E - U(x)\right]}, \qquad (3.30)$$

where the sign of $\dot{x}$ is determined from the initial conditions.

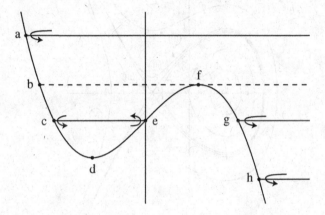

Fig. 3.1   Bounded and unbounded energy levels in a cubic potential $U(x) = x - x^3/3$.

It is immediately clear that physical motion is possible only if $E \geq U(x)$; points where $E = U(x)$ are known as *turning* points since the particle velocity $\dot{x}$ vanishes at these points. In Fig. 3.1, which represents the dimensionless potential $U(x) = x - x^3/3$, each horizontal line corresponds to a constant energy value (called an energy *level*). For the top energy level, only one turning point (labeled a in Fig. 3.1) exists and a particle coming from the right will be *reflected* at point a and return to large (positive) values of x; the motion in this case is said to be along an *unbounded* orbit (see orbits I in Fig. 3.2). As the energy value is lowered, two turning points (labeled b and f) appear and motion can either be *bounded* (between points b and f) or unbounded (if the initial position is to the right of point f); this energy level is known as a *separatrix* level since bounded and unbounded motions share one turning point (see orbits II and III in Fig. 3.2). As energy is lowered below the separatrix level, three turning points (labeled c, e, and g) appear and, once again, motion can either be along a *bounded* orbit (with turning points c and e) or an unbounded orbit if the initial position is to the right of point g (see orbits IV and V in Fig. 3.2).[1] Lastly, we note that point d in Fig. 3.1 is actually an equilibrium point (as is point f), where $\dot{x}$ and $\ddot{x}$ both vanish; only unbounded motion is allowed as energy is lowered below point d (e.g., point h) and the corresponding unbounded orbits are analogous to orbit V in Fig. 3.2.

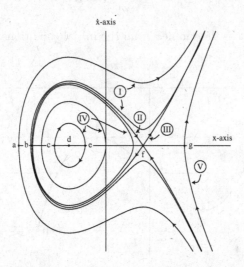

Fig. 3.2   Bounded and unbounded orbits in the cubic potential shown in Fig. 3.1: the orbits I correspond to the energy level with turning point a; the bounded orbit II and unbounded orbit III correspond to the separatrix energy level with turning points b and f; the bounded orbits IV correspond to the energy level with turning points c and e; and the unbounded orbit V corresponds to the energy level with turning point g. (These orbits are explicitly solved in Appendix B in terms of the Weierstrass elliptic function.)

The dynamical solution $x(t; E)$ of the Hamilton's equations (3.29) is first expressed an integration by quadrature using Eq. (3.30) as

$$t(x; E) = \sqrt{\frac{m}{2}} \int_{x_0}^{x} \frac{ds}{\sqrt{E - U(s)}}, \tag{3.31}$$

where the particle's initial position x_0 is between the turning points $x_1 < x_2$ (allowing $x_2 \to \infty$) and we assume that $\dot{x}(0) > 0$. Next, inversion of the relation (3.31) yields the solution $x(t; E)$.

Lastly, for bounded motion in one dimension, the particle bounces back and forth between the two turning points x_1 and $x_2 > x_1$, and the period

---

[1]Note: Quantum tunneling establishes a connection between the bounded and unbounded solutions separated by unphysical regions (where $E < U$).

of oscillation $T(E)$ is a function of energy alone

$$T(E) = 2 \int_{x_1}^{x_2} \frac{dx}{\dot{x}(x, E)} = \sqrt{2m} \int_{x_1}^{x_2} \frac{dx}{\sqrt{E - U(x)}}. \tag{3.32}$$

Thus, Eqs. (3.31) and (3.32) describe applications of the Energy Method in one dimension. We now look at a series of one-dimensional problems solvable by the Energy Method.

3.5.2 *Simple Harmonic Oscillator*

As a first example, we consider the case of a particle of mass m attached to a spring of constant k, for which the potential energy is $U(x) = \frac{1}{2} kx^2$. The motion of a particle with total energy E is always bounded, with turning points

$$x_{1,2}(E) = \pm \sqrt{2E/k} = \pm a.$$

We start with the solution (3.31) for $t(x; E)$ for the case of $x(0; E) = +a$, so that $\dot{x}(t; E) < 0$ for $t > 0$, and

$$t(x; E) = \sqrt{\frac{m}{k}} \int_x^a \frac{ds}{\sqrt{a^2 - s^2}} = \sqrt{\frac{m}{k}} \arccos\left(\frac{x}{a}\right). \tag{3.33}$$

Inversion of this relation yields the well-known solution $x(t; E) = a \cos(\omega_0 t)$, where $\omega_0 = \sqrt{k/m}$. Next, using Eq. (3.32), we find the period of oscillation

$$T(E) = \frac{4}{\omega_0} \int_0^a \frac{dx}{\sqrt{a^2 - x^2}} = \frac{2\pi}{\omega_0},$$

which turns out to be independent of energy E.

3.5.3 *Pendulum*

Our second example involves the case of the pendulum of length ℓ and mass m in a gravitational field g (see Sec. 2.4.1). The energy equation in this case is

$$E = \frac{1}{2} m\ell^2 \dot{\theta}^2 + mg\ell (1 - \cos\theta). \tag{3.34}$$

The total energy of the pendulum is determined from its initial conditions $(\theta_0, \dot{\theta}_0)$:

$$E = \frac{1}{2} m\ell^2 \dot{\theta}_0^2 + mg\ell (1 - \cos\theta_0),$$

where the potential energy term is $mg\ell(1 - \cos\theta) \leq 2mg\ell$. Solutions of the pendulum problem (3.34) are divided into three classes depending on the value of the total energy of the pendulum (see Fig. 3.3): Class III (rotation) $E > 2mg\ell$, Class II (separatrix) $E = 2mg\ell$, and Class I (libration) $E < 2mg\ell$.

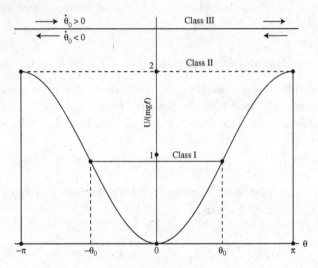

Fig. 3.3   Normalized pendulum potential $U(\theta)/(mg\ell) = 1 - \cos\theta$.

In the rotation class ($E > 2mg\ell$), the kinetic energy can never vanish and the pendulum keeps rotating either clockwise or counter-clockwise depending on the sign of $\dot{\theta}_0$. In the libration class ($E < 2mg\ell$), on the other hand, the kinetic energy vanishes at turning points easily determined by initial conditions if the pendulum starts from rest ($\dot{\theta}_0 = 0$) – in this case, the turning points are $\pm\theta_0$, where

$$\theta_0 = \arccos\left(1 - \frac{E}{mg\ell}\right).$$

In the separatrix class ($E = 2mg\ell$), the turning points are $\theta_0 = \pm\pi$. The numerical solution of the normalized pendulum equation $\theta'' + \sin\theta = 0$ subject to the initial condition θ_0 and $\theta_0' = \pm\sqrt{2(\epsilon - 1 + \cos\theta_0)}$ yields the following curves (see Fig. 3.3). Here, the three classes I, II, and III are easily seen (with $\epsilon = 1 - \cos\theta_0$ and $\theta_0' = 0$ for classes I and II and $\epsilon > 1 - \cos\theta_0$ for class III). Note that for rotations (class III), the pendulum slows down as it approaches $\theta = \pm\pi$ (the top part of the circle) and speeds up as it approaches $\theta = 0$ (the bottom part of the circle). In fact, since $\theta = \pi$

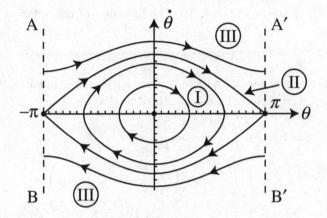

Fig. 3.4 Phase portrait of the pendulum.

and $\theta = -\pi$ represent the same point in space, the lines AB and $A'B'$ in Fig. 3.4 should be viewed as being identical (i.e., they should be glued together) and the geometry of the *phase space* for the pendulum problem is actually that of a cylinder.

3.5.3.1 *Libration Class* $(E < 2mg\ell)$

We now look at an explicit solution for pendulum librations (class I), where the angular velocity $\dot\theta$ is

$$\dot\theta(\theta; E) = \pm\omega_0 \sqrt{2(\cos\theta - \cos\theta_0)} = \pm 2\omega_0 \sqrt{\sin^2(\theta_0/2) - \sin^2(\theta/2)}, \tag{3.35}$$

where $\omega_0 = \sqrt{g/\ell}$ and $\pm\theta_0$ are the turning points for this problem. By making the substitution $\sin\theta/2 = k\,\sin\varphi$, where

$$k(E) = \sin[\theta_0(E)/2] = \sqrt{\frac{E}{2\,mg\ell}} < 1 \tag{3.36}$$

and $\varphi = \pm\pi/2$ when $\theta = \pm\theta_0$, Eq. (3.35) becomes

$$\dot\varphi = \pm\omega_0 \sqrt{1 - k^2 \sin^2\varphi}. \tag{3.37}$$

The libration solution of the pendulum problem is thus

$$\omega_0\,t(\theta; E) \equiv \tau(\theta; E) = \int_{\Theta(\theta;E)}^{\pi/2} \frac{d\varphi}{\sqrt{1 - k^2 \sin^2\varphi}}, \tag{3.38}$$

where $\Theta(\theta; E) = \arcsin(k^{-1} \sin \theta/2)$. The inversion of this relation yields the solution

$$\theta(t; E) = 2 \arcsin \left[k \, \text{sn}(\tau | k^2) \right] \tag{3.39}$$

expressed in terms of the Jacobi elliptic function sn (see Appendix B.1 for more details). The period of oscillation is defined as

$$\begin{aligned}
\omega_0 \, T(E) &= 4 \int_0^{\pi/2} \frac{d\varphi}{\sqrt{1 - k^2 \sin^2 \varphi}} \\
&= 4 \int_0^{\pi/2} d\varphi \left(1 + \frac{k^2}{2} \sin^2 \varphi + \cdots \right), \\
&= 2\pi \left(1 + \frac{k^2}{4} + \cdots \right) = 4 \, K(k^2),
\end{aligned} \tag{3.40}$$

where $K(k)$ denotes the complete elliptic integral of the first kind (see Fig. 3.5 and Appendix B.1).

We note here that if $k \ll 1$ (or $\theta_0 \ll 1$) the libration period of a pendulum is nearly independent of energy, $T \simeq 2\pi/\omega_0$. However, we also note that as $E \to 2mg\ell$ ($k \to 1$ or $\theta_0 \to \pi$), the libration period of the pendulum becomes infinitely large, i.e., $T \to \infty$ in Eq. (3.40) (see Fig. 3.5).

3.5.3.2 *Separatrix Class* $(E = 2mg\ell)$

In the separatrix case ($\theta_0 = \pi$), the pendulum equation (3.35) yields the *separatrix* equation $\dot{\varphi} = \omega_0 \cos \varphi$, where $\varphi = \theta/2$. The separatrix solution is expressed in terms of the transcendental equation

$$\sec \varphi(t) = \cosh(\tau + \gamma), \tag{3.41}$$

where $\cosh \gamma = \sec \varphi_0$ represents the initial condition. We again note that $\varphi \to \pi/2$ (or $\theta \to \pi$) only as $t \to \infty$. Separatrices are associated with turning points s_0 where $U'(s_0) = 0$ and $U''(s_0) < 0$, which are quite common in periodic dynamical systems involving bounded and unbounded orbits (see also Sec. 7.2.3).

3.5.3.3 *Rotation Class* $(E > 2mg\ell)$

The solution for rotations (class III) associated with the initial conditions $\theta_0 = \pm\pi$ and

$$\frac{1}{2} \dot{\theta}_0^2 = \omega_0^2 \left(\frac{E}{mg\ell} - 2 \right) = \frac{1}{2} \dot{\theta}^2 - \omega_0^2 \left(1 + \cos \theta \right),$$

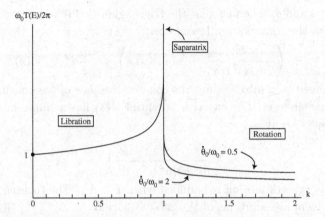

Fig. 3.5 Normalized pendulum period $\omega_0 T(E)/2\pi$ as a function of the normalized energy $k^2 = E/(2mg\ell)$ for Libration Class I ($k^2 < 1$) and Rotation Class III ($k^2 > 1$). The period is infinite for the Separatrix Class II ($k^2 = 1$).

or

$$\dot{\theta} = \pm \sqrt{\dot{\theta}_0^2 + 2\omega_0^2 (1 + \cos\theta)},$$

which shows that $\dot{\theta}$ does not vanish for rotations. We now write $\cos\theta = 1 - 2\sin^2(\theta/2)$ and define $\varphi \equiv \theta/2$ to obtain

$$\dot{\varphi} = \pm k\omega_0 \sqrt{1 - k^{-2}\sin^2\varphi}, \qquad (3.42)$$

where $k(E)$ is defined in Eq. (3.36). Hence, the solution for rotations is expressed in terms of the Jacobi elliptic function sn (see Appendix B.1 for more details) as

$$\theta(t; E) = 2\arcsin\left[\operatorname{sn}(k\,\tau|k^{-2})\right] \qquad (3.43)$$

The rotation period is defined as

$$\omega_0 T(E) = \frac{4}{k} \int_0^{\pi/2} \frac{d\varphi}{\sqrt{1 - k^{-2}\sin^2\varphi}} = \frac{4}{k} K(k^{-2}).$$

Figure 3.5 shows the plot of the normalized pendulum period as a function of k^2 for the Libration Class I and Rotation Class III (two cases are shown: $\dot{\theta}_0/\omega_0 \equiv 2\sqrt{k^2 - 1} = 2$ and 0.5).

3.5.4 *Constrained Motion on the Surface of a Cone*

The constrained motion of a particle of mass m on a cone in the presence of gravity was shown in Sec. 2.5.4 to be doubly periodic in the generalized coordinates s and θ. The fact that the Lagrangian (2.46) is independent of time leads to the conservation law of energy

$$E = \frac{m}{2}\dot{s}^2 + \left(\frac{\ell^2}{2m\sin^2\alpha\, s^2} + mg\cos\alpha\, s\right) = \frac{m}{2}\dot{s}^2 + V(s), \quad (3.44)$$

where we have taken into account the conservation law of angular momentum $\ell = ms^2\sin^2\alpha\,\dot{\theta}$. The effective potential $V(s)$ has a single minimum $V_0 = \frac{3}{2}mgs_0\cos\alpha$ at

$$s_0 = \left(\frac{\ell^2}{m^2 g\,\sin^2\alpha\,\cos\alpha}\right)^{\frac{1}{3}},$$

and the only type of motion is bounded when $E > V_0$. The turning points for this problem are solutions of the cubic equation

$$\frac{3}{2}\epsilon = \frac{1}{2\sigma^2} + \sigma,$$

where $\epsilon = E/V_0$ and $\sigma = s/s_0$. One of the three roots remains negative for all normalized energies ϵ; this root is unphysical since s must be positive (by definition). The other two roots (σ_1, σ_2), which are complex for $\epsilon < 1$ (i.e., for energies below the minimum of the effective potential energy V_0), become real at $\epsilon = 1$, where $\sigma_1 = \sigma_2$, and separate $(\sigma_1 < \sigma_2)$ for larger values of ϵ. Lastly, by defining the frequency $\omega_g \equiv \sqrt{(g/s_0)}\cos\alpha$, the period of radial oscillations is determined by the definite integral

$$\omega_g T(\epsilon) = 2\int_{\sigma_1}^{\sigma_2} \frac{\sigma\, d\sigma}{\sqrt{3\epsilon\,\sigma^2 - 1 - 2\sigma^3}},$$

whose solution is expressed in terms of Weierstrass elliptic function (see Appendix B). Lastly, we note that in one period (see Fig. 2.9), the orbit precesses by an angular deviation

$$\Delta\theta = \int_0^{\omega_g T} \theta'\, d\tau = \frac{1}{\sin\alpha}\int_0^{\omega_g T} \frac{d\tau}{\sigma^2},$$

where $\theta' = d\theta/d\tau \equiv \omega_g^{-1}\dot{\theta}$.

3.6 Problems

1. A particle of mass m and total energy E moves periodically in a one-dimensional potential $U(x) = F|x|$, where F is a positive constant.

(a) Find the turning points for this potential.

(b) Find the dynamical solution $x(t; E)$ for this potential by choosing a suitable initial condition.

(c) Find the period $T(E)$ for the motion.

2. A block of mass m rests on the inclined plane (with angle θ) of a triangular block of mass M as shown in Fig. 3.6. Here, we consider the case where both blocks slide without friction (i.e., m slides on the inclined plane without friction and M slides without friction on the horizontal plane).

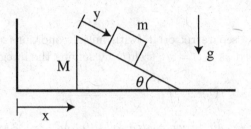

Fig. 3.6   Problem 2.

(a) Using the generalized coordinates (x, y) shown in Fig. 3.6, construct the Lagrangian $L(x, \dot{x}, y, \dot{y})$.

(b) Derive the Euler-Lagrange equations for x and y.

(c) Calculate the canonical momenta

$$p_x(x, \dot{x}, y, \dot{y}) = \frac{\partial L}{\partial \dot{x}} \quad \text{and} \quad p_y(x, \dot{x}, y, \dot{y}) = \frac{\partial L}{\partial \dot{y}},$$

and invert these expressions to find the functions $\dot{x}(x, p_x, y, p_y)$ and $\dot{y}(x, p_x, y, p_y)$.

(d) Calculate the Hamiltonian $H(x, p_x, y, p_y)$ for this system by using the Legendre transformation $H(x, p_x, y, p_y) = p_x \dot{x} + p_y \dot{y} - L(x, \dot{x}, y, \dot{y})$, where the functions $\dot{x}(x, p_x, y, p_y)$ and $\dot{y}(x, p_x, y, p_y)$ are used.

(e) Find which of the two momenta found in Part (c) is a constant of the motion and discuss why it is so. If the two blocks start from rest, what is the value of this constant of motion?

3. Consider all possible orbits of a unit-mass particle moving in the dimensionless potential $U(x) = 1 - x^2/2 + x^4/16$. Here, orbits are solutions of the equation of motion $\ddot{x} = -U'(x)$ and the dimensionless energy equation is $E = \dot{x}^2/2 + U(x)$.

(a) Draw the potential $U(x)$ and identify all possible unbounded and bounded orbits (with their respective energy ranges).

(b) For each orbit found in part (a), find the turning point(s) for each energy level.

(c) Sketch the phase portrait $(x, \dot{x})$ showing all orbits (including the separatrix orbit).

(d) Show that the separatrix orbit (with initial conditions $x_0 = \sqrt{8}$ and $\dot{x}_0 = 0$) is expressed as $x(t) = \sqrt{8}\,\text{sech}(t)$ by solving the integral

$$t(x) = \int_x^{\sqrt{8}} \frac{ds}{\sqrt{s^2\,(1 - s^2/8)}}.$$

(*Hint: use the hyperbolic trigonometric substitution $s = \sqrt{8}\,\text{sech}\,\xi$.*)

4. Write a numerical code to solve the second-order ordinary differential equation $\ddot{x} = x - x^3/3$ by choosing appropriate initial conditions needed to obtain all the possible (bounded and unbounded) orbits.

5. When a particle (of mass m) moving under the potential $U(x)$ is perturbed by the potential $\delta U(x)$, its period (3.32) is changed by a small amount defined as

$$\delta T = -\sqrt{2m}\,\frac{\partial}{\partial E}\left[\int_{x_1}^{x_2} \frac{\delta U(x)\,dx}{\sqrt{E - U(x)}}\right],$$

where $x_{1,2}$ are the turning points of the unperturbed problem. Calculate the change in the period of a particle moving in the quadratic potential $U(x) = m\omega^2\,x^2/2$ introduced by the perturbation potential $\delta U(x) = \epsilon x^4$. Here, ω denotes the unperturbed frequency, the particle is trapped in the region $-a \leq x \leq a$, and ϵ is a constant that satisfies the condition $\epsilon \ll m\omega^2/(2a^2)$.

6. Consider two simple-harmonic oscillators described by the coordinates x_1 and x_2 with respective frequencies ω_1 and $\omega_2 > \omega_1$ and subject to the (nonlinear) constraint $x_1^2 + x_2^2 = 1$.

(a) Derive the Euler-Lagrange equations for x_1 and x_2 by using the Lagrangian

$$L = \frac{1}{2}\left(\dot{x}_1^2 + \dot{x}_2^2\right) - \frac{1}{2}\left(\omega_1^2\, x_1^2 + \omega_2^2\, x_2^2\right) - \lambda\left(x_1^2 + x_2^2 - 1\right),$$

where λ is the Lagrange multiplier associated with the constraint, and verify that the energy $E = \frac{1}{2}\left(\dot{x}_1^2 + \dot{x}_2^2\right) + \frac{1}{2}\left(\omega_1^2\, x_1^2 + \omega_2^2\, x_2^2\right)$ is a constant of the motion (i.e., $dE/dt = 0$).

(b) Find an expression for the Lagrange multiplier λ in terms of x_1 and x_2.

(c) By substituting $x_1(t) = \cos\theta(t)$ and $x_2(t) = \sin\theta(t)$ into the Euler-Lagrange equations derived in Part (a), show that one obtains the nonlinear equation

$$\ddot{\theta} + \left(\omega_2^2 - \omega_1^2\right)\cos\theta\,\sin\theta = 0.$$

Note that this equation can also be written as $\varphi'' + (\Omega^2 - 1)\sin\varphi = 0$, where $\varphi(\tau) = 2\,\theta(\omega_1\, t)$ and $\Omega \equiv \omega_2/\omega_1 > 1$.

7. A cycloidal pendulum (of mass m) is suspended from the cusp of a cycloid cut in a rigid support. The path described by the pendulum is cycloidal and is given by the parameterized equations $x = a\,(\phi - \sin\phi)$ and $y = a\,(\cos\phi - 1)$, where the length of the pendulum is $\ell = 4a$ and $0 \le \phi \le 2\pi$.

(a) Construct the Lagrangian $L(\phi, \dot{\phi})$ and show that the equation of motion of the pendulum is

$$2\,\ddot{\phi}\,(1 - \cos\phi) + \sin\phi\left(\dot{\phi}^2 - 4\omega_g^2\right) = 0, \qquad (3.45)$$

where $\omega_g = \sqrt{g/\ell}$.

(b) Construct the total energy as a function of ϕ and $\dot{\phi}$.

(c) Show that Eq. (3.45) can be written as $\ddot{u} + \omega_0^2\, u = 0$, where $u = \cos(\phi/2)$. Hence, the cycloidal pendulum has a period $2\pi/\omega_0$.

8. Recreate the phase portrait shown in Fig. 3.4 for the pendulum by numerical integration of the normalized equation $\theta'' + \sin\theta = 0$ subject to the energy conservation law $\epsilon = \frac{1}{2}(\theta')^2 + (1 - \cos\theta)$, from which suitable initial conditions can be selected.

9. Show that Eq. (3.41) is the separatrix solution for the pendulum problem when $E = 2\,mg\,\ell$.

10. A particle of mass m is moving in the potential

$$U(x, y) = \begin{cases} U_1 & (x < 0) \\ U_2 & (x > 0) \end{cases}$$

(a) By using all known conservation laws for this problem, show that if the particle is moving with velocity $\mathbf{v}_1 = v_1\,(\cos\theta_1\,\hat{\mathbf{x}} + \sin\theta_1\,\hat{\mathbf{y}})$ in the region $x < 0$, then the particle is moving along the direction $\theta_2 \neq \theta_1$ in the region $x > 0$, where

$$\frac{\sin\theta_1}{\sin\theta_2} = \sqrt{1 + \frac{U_1 - U_2}{E - U_1}}.$$

(b) Discuss the cases $E > U_1 > U_2$ and $E > U_2 > U_1$.

11. Show that the period of oscillation for a particle of mass m moving in the potential $U(x) = U_0 \tan^2(kx)$ is given as

$$T(E) = \frac{\pi}{k}\sqrt{\frac{2\,m}{E + U_0}}.$$

Hint: Use the integral

$$\int_0^{\pi/2} \frac{d\theta}{1 + \alpha^2 \sin^2\theta} = \frac{\pi/2}{\sqrt{1 + \alpha^2}}.$$

12. The relativistic Lagrangian for a particle of rest mass m moving in a potential $U(\mathbf{x})$ is

$$L(\mathbf{x}, \dot{\mathbf{x}}) = -mc^2\,\sqrt{1 - |\dot{\mathbf{x}}|^2/c^2} - U(\mathbf{x}),$$

where c is the speed of light.

(a) Derive the equation of motion for $\mathbf{x}$.

(b) Using the Legendre transformation (3.4), derive the relativistic Hamiltonian.

Chapter 4

Motion in a Central-Force Field

The present Chapter introduces an important set of problems that are solvable by the Energy and Noether methods. Here, bounded and unbounded solutions are obtained for (time-independent) central-force planar problems in which energy and angular momentum are constants of the motion. The existence of two constants of motion for two-dimensional planar motion implies that exact solutions can be obtained if integral solutions (obtained by quadrature) can be inverted.

4.1 Motion in a Central-Force Field

A particle moves under the influence of a central-force field

$$\mathbf{F}(\mathbf{r}) \;=\; F(r)\,\widehat{r}(\theta,\varphi) \;\equiv\; -\,U'(r)\,\widehat{r}, \tag{4.1}$$

if the force $F(r) = -U'(r)$ on the particle is independent of the angular position (θ, φ) of the particle about the center of force and depends only on its distance r from the center of force. Note that, for a central-force potential $U(r)$, the angular momentum $\mathbf{L} = \ell\widehat{z}$ in the CM frame is a constant of the motion since $\mathbf{r} \times \nabla U(r) = 0$ [see Eq. (2.54)].

4.1.1 *Lagrangian Formalism*

The motion of two particles in an isolated system takes place on a two-dimensional plane, which we henceforth take to be the (x, y)-plane and, hence, the constant angular momentum is $\mathbf{L} = \ell\widehat{z}$. When these particles move in a central-force field (4.1), the center-of-mass Lagrangian is simply

$$L \;=\; \frac{\mu}{2}\left(\dot{r}^2 + r^2\,\dot{\theta}^2\right) - U(r), \tag{4.2}$$

95

where μ denotes the reduced mass for the two-particle system and polar coordinates (r, θ) are most conveniently used, with $x = r \cos \theta$ and $y = r \sin \theta$. Since the potential U is independent of θ, it follows from Noether's Theorem that the canonical angular momentum

$$p_\theta = \frac{\partial L}{\partial \dot{\theta}} = \mu r^2 \dot{\theta} \equiv \ell \qquad (4.3)$$

is a constant of motion (labeled ℓ). The Euler-Lagrange equation for r, therefore, becomes the radial force equation

$$\mu \left(\ddot{r} - r \dot{\theta}^2 \right) = \mu \ddot{r} - \frac{\ell^2}{\mu r^3} = F(r) \equiv - U'(r). \qquad (4.4)$$

In this description, the planar orbit is parameterized by time, i.e., once $r(t)$ and $\theta(t)$ are obtained, a path $r(\theta)$ onto the plane is defined.

4.1.1.1 Radial Orbit Equation

Since $\dot{\theta} = \ell / \mu r^2$ does not change sign along the orbit (as a result of the conservation of angular momentum $\ell \neq 0$), we may replace $\dot{r}$ and $\ddot{r}$ with $r'(\theta)$ and $r''(\theta)$ as follows. First, we begin with

$$\dot{r} = \dot{\theta} r' = \frac{\ell r'}{\mu r^2} = - \frac{\ell}{\mu} \left(\frac{1}{r} \right)' = - (\ell/\mu) s',$$

where we use the conservation of angular momentum and define the new dependent variable $s(\theta) = 1/r(\theta)$. Next, we write $\ddot{r} = - (\ell/\mu) \dot{\theta} s'' = - (\ell/\mu)^2 s^2 s''$, so that the radial force equation (4.4) becomes

$$s'' + s = - \frac{\mu}{\ell^2 s^2} F(1/s) \equiv - \frac{d\overline{U}(s)}{ds}, \qquad (4.5)$$

where

$$\overline{U}(s) = \frac{\mu}{\ell^2} U(1/s) \qquad (4.6)$$

denotes the normalized central potential expressed as a function of s.

4.1.1.2 Inversion Problem

Note that the form of the potential $U(r)$ can be calculated from the function $s(\theta) = 1/r(\theta)$. For example, consider the particle trajectory described in terms of the function $r(\theta) = r_0 \sec(\alpha \theta)$, where r_0 and α are constants. The radial orbit equation (4.5) then becomes

$$s'' + s = - (\alpha^2 - 1) s = - \frac{d\overline{U}(s)}{ds},$$

and thus

$$\overline{U}(s) \;=\; \frac{1}{2}\,(\alpha^2 - 1)\,s^2 \;\;\rightarrow\;\; U(r) \;=\; \frac{\ell^2}{2\mu\,r^2}\,(\alpha^2 - 1).$$

As expected, the central potential is either repulsive for $\alpha > 1$ or attractive for $\alpha < 1$ (see Fig. 4.1).

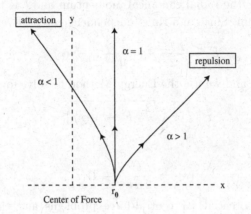

Fig. 4.1   Repulsive ($\alpha > 1$) and attractive ($\alpha < 1$) orbits for the central-force potential $U(r) = (\ell^2/2\mu)\,(\alpha^2 - 1)\,r^{-2}$.

The function $\theta(t)$ is determined from the relation

$$\dot{\theta} \;=\; \frac{\ell}{\mu r^2(\theta)} \;\;\rightarrow\;\; t(\theta) \;=\; \frac{\mu}{\ell}\int_0^\theta r^2(\phi)\,d\phi.$$

Hence, we find

$$t(\theta) \;=\; \frac{\mu r_0^2}{\alpha\ell}\int_0^{\alpha\theta} \sec^2\phi\,d\phi \;=\; \frac{\mu r_0^2}{\alpha\ell}\,\tan(\alpha\theta).$$

Upon substituting $r(\theta) = r_0\,\sec(\alpha\,\theta)$, we obtain

$$t(r) \;=\; \frac{\mu r_0^2}{\alpha\,\ell}\,\sqrt{\left(\frac{r}{r_0}\right)^2 - 1},$$

which yields the solution

$$r(t) \;=\; r_0\,\sqrt{1 + \left(\frac{\alpha\ell\,t}{\mu r_0^2}\right)^2}$$

and the total energy

$$E \;=\; \frac{\alpha^2\,\ell^2}{2\mu r_0^2},$$

is determined from the initial conditions $r(0) = r_0$ and $\dot{r}(0) = 0$.

4.1.2 *Hamiltonian Formalism*

The Hamiltonian for the central-force problem (4.2) is

$$H = \frac{p_r^2}{2\mu} + \frac{\ell^2}{2\mu r^2} + U(r),$$

where $p_r = \mu \dot{r}$ is the radial canonical momentum and ℓ is the conserved angular momentum. Since in a time-independent potential, energy

$$E = \frac{\mu \dot{r}^2}{2} + \frac{\ell^2}{2\mu r^2} + U(r)$$

is also conserved, and we use the Energy Method to solve for $\dot{r}(r; E, \ell)$ as

$$\dot{r} = \pm \sqrt{\frac{2}{\mu} [\, E - V(r)\,]}, \tag{4.7}$$

where

$$V(r) = \frac{\ell^2}{2\mu r^2} + U(r) \tag{4.8}$$

is the *effective* potential for central-force problems and the sign $\pm$ in Eq. (4.7) depends on initial conditions. Hence, Eq. (4.7) yields the integral solution

$$t(r; E, \ell) = \sqrt{\frac{\mu}{2}} \int_{r_0}^{r} \frac{ds}{\sqrt{E - V(s)}}, \tag{4.9}$$

which is identical in form with the integral solution (3.31) for one-dimensional problems.

When we want to determine the orbit $r(\theta)$, we can also use the energy equation

$$\mu E/\ell^2 \equiv \frac{\epsilon}{2} = \frac{s'^2}{2} + \frac{s^2}{2} + \overline{U}(s), \tag{4.10}$$

to obtain

$$s'(\theta) = \pm \sqrt{\epsilon - 2\overline{U}(s) - s^2}. \tag{4.11}$$

Hence, for a given central-force potential $U(r)$, we can solve for $r(\theta) = 1/s(\theta)$ by integrating

$$\theta(s) = - \int_{s_0}^{s} \frac{d\sigma}{\sqrt{\epsilon - 2\overline{U}(\sigma) - \sigma^2}}, \tag{4.12}$$

where s_0 defines $\theta(s_0) = 0$, and performing the inversion $\theta(s) \rightarrow s(\theta) = 1/r(\theta)$.

4.1.3 Turning Points

Equation (4.11) yields the following energy equation

$$E = \frac{\mu}{2}\dot{r}^2 + \frac{\ell^2}{2\mu r^2} + U(r) = \frac{\ell^2}{2\mu}\left[(s')^2 + s^2 + 2\overline{U}(s)\right],$$

where $s' = -\mu\dot{r}/\ell$. *Turning* points are those special values of r_n (or s_n) ($n = 1, 2, ...$) for which

$$E = U(r_n) + \frac{\ell^2}{2\mu r_n^2} = \frac{\ell^2}{\mu}\left[\overline{U}(s_n) + \frac{s_n^2}{2}\right],$$

i.e., $\dot{r}$ (or s') vanishes at these points. If two non-vanishing turning points $r_2 < r_1 < \infty$ (or $0 < s_1 < s_2$) exist, the motion is said to be *bounded* in the interval $r_2 < r < r_1$ (or $s_1 < s < s_2$), otherwise the motion is *unbounded*. If the motion is bounded, the angular period $\Delta\theta$ is defined as

$$\Delta\theta = 2\int_{s_1}^{s_2} \frac{ds}{\sqrt{\epsilon - 2\overline{U}(s) - s^2}}. \tag{4.13}$$

Here, the bounded orbit is *closed* only if $\Delta\theta$ is a rational multiple of 2π.

4.2 Homogeneous Central Potentials*

An important class of central potentials is provided by homogeneous potentials that satisfy the condition $U(\lambda\mathbf{r}) = \lambda^n U(\mathbf{r})$, where λ denotes a rescaling parameter and n denotes the *order* of the homogeneous potential.

4.2.1 The Virial Theorem

The Virial Theorem is an important theorem in Celestial Mechanics and Astrophysics. We begin with the time derivative of the quantity $S = \sum_i \mathbf{p}_i \cdot \mathbf{r}_i$:

$$\frac{dS}{dt} = \sum_i \left(\frac{d\mathbf{p}_i}{dt}\cdot\mathbf{r}_i + \mathbf{p}_i\cdot\frac{d\mathbf{r}_i}{dt}\right), \tag{4.14}$$

where the summation is over all particles in a mechanical system under the influence of a self-interaction potential

$$U = \frac{1}{2}\sum_{i,j\neq i} U(\mathbf{r}_i - \mathbf{r}_j) \equiv \frac{1}{2}\sum_{i,j\neq i} U_{ij}.$$

We note, however, that S itself can be written as a time derivative

$$S = \sum_i m_i \frac{d\mathbf{r}_i}{dt} \cdot \mathbf{r}_i = \frac{d}{dt}\left(\frac{1}{2}\sum_i m_i |\mathbf{r}_i|^2\right) = \frac{1}{2}\frac{d\mathcal{I}}{dt},$$

where $\mathcal{I}$ denotes the *moment of inertia* of the system. Using Hamilton's equations

$$\frac{d\mathbf{r}_i}{dt} = \frac{\mathbf{p}_i}{m_i} \quad \text{and} \quad \frac{d\mathbf{p}_i}{dt} = -\sum_{j \neq i} \nabla_i U(\mathbf{r}_i - \mathbf{r}_j),$$

Eq. (4.14) can also be written as

$$\frac{1}{2}\frac{d^2\mathcal{I}}{dt^2} = \sum_i \left(\frac{|\mathbf{p}_i|^2}{m_i} - \mathbf{r}_i \cdot \sum_{j \neq i} \nabla_i U_{ij}\right) = 2\,K - \sum_{i,\,j \neq i} \mathbf{r}_i \cdot \nabla_i U_{ij},$$

$$(4.15)$$

where K denotes the kinetic energy of the mechanical system. Next, using Newton's Third Law, we write

$$\sum_{i,\,j \neq i} \mathbf{r}_i \cdot \nabla_i U_{ij} = \frac{1}{2}\sum_{i,\,j \neq i} (\mathbf{r}_i - \mathbf{r}_j) \cdot \nabla U(\mathbf{r}_i - \mathbf{r}_j),$$

and, for a homogeneous central potential of order n, we find $\mathbf{r} \cdot \nabla U(\mathbf{r}) = n\,U(\mathbf{r})$, so that

$$\frac{1}{2}\sum_{i,\,j \neq i} (\mathbf{r}_i - \mathbf{r}_j) \cdot \nabla U(\mathbf{r}_i - \mathbf{r}_j) \equiv n\,U.$$

Hence, Eq. (4.15) becomes the *Virial of Clausius* (Rudolph Clausius, 1822-1888)

$$\frac{1}{2}\frac{d^2\mathcal{I}}{dt^2} = 2\,K - n\,U. \tag{4.16}$$

If we now assume that the mechanical system under consideration is periodic in time (e.g., the system is bounded), then the time average (denoted $\langle \cdots \rangle$) of Eq. (4.16) yields the Virial Theorem

$$\langle K \rangle = \frac{n}{2}\langle U \rangle, \tag{4.17}$$

so that the time-average of the total energy of the mechanical system, $E = K + U$, is expressed as

$$E = (1 + n/2)\langle U \rangle = (1 + 2/n)\langle K \rangle,$$

since $\langle E \rangle = E$. For example, for the Kepler problem ($n = -1$), we find

$$E = \frac{1}{2}\langle U \rangle = -\langle K \rangle < 0, \tag{4.18}$$

which means that the total energy of a bounded Keplerian orbit is negative (see Sec. 4.3.1). We note that the Virial Theorem has important applications in astrophysics where the contraction of a self-gravitating cloud (i.e., $\langle U \rangle$ becoming more negative) leads to an increase in its internal energy (i.e., $\langle K \rangle$ becoming more positive).

4.2.2 General Properties of Homogeneous Potentials

We now investigate the dynamical properties of orbits in homogeneous central potentials of the form $U(r) = (k/n)\, r^n$ ($n \neq -2$), where k denotes a positive constant. Note that the central force $\mathbf{F} = -\nabla U = -k\, r^{n-1}\hat{r}$ is attractive if $k > 0$.

First, the effective potential (4.8) has an extremum at a distance $r_0 = 1/s_0$ defined as

$$r_0^{n+2} = \frac{\ell^2}{k\mu} = \frac{1}{s_0^{n+2}}.$$

It is simple to show that this extremum is a maximum if $n < -2$ or a minimum if $n > -2$; we shall, henceforth, focus our attention on the latter case, where the minimum in the effective potential is

$$V_0 = V(r_0) = \left(1 + \frac{n}{2}\right)\frac{k}{n}\, r_0^n = \left(1 + \frac{n}{2}\right) U_0.$$

In the vicinity of this minimum, we can certainly find periodic orbits with turning points ($r_2 = 1/s_2 < r_1 = 1/s_1$) that satisfy the condition $E = V(r)$.

Next, the radial equation (4.5) is written in terms of the potential $\overline{U}(s) = (\mu/\ell^2)\, U(1/s)$ as

$$s'' + s = -\frac{d\overline{U}}{ds} = \frac{s_0^{n+2}}{s^{n+1}},$$

and its solution is given as the orbit integral

$$\theta(s) = \int_s^{s_2} \frac{d\sigma}{\sqrt{\epsilon - (2/n)\, s_0^{n+2}/\sigma^n - \sigma^2}}, \tag{4.19}$$

where s_2 denotes the upper turning point in the s-coordinate. The solution (4.19) can be expressed in terms of closed analytic expressions obtained by trigonometric substitution only for $n = -1$ or $n = 2$ (when $\epsilon \neq 0$), which we now study in detail below (the cases $n = -3$ and -4, for example, are solved in terms of elliptic functions as discussed in Appendix B).

4.3 Kepler Problem

In this Section, we solve the Kepler problem (Johannes Kepler, 1571-1630) where the central potential $U(r) = -k/r$ is homogeneous with order $n = -1$ and k is a positive constant.[1] The Virial Theorem (4.17) implies that

---

[1] For problems involving gravitational attraction, we have $k = G\, m_1 m_2$, while for problems involving electrostatic attraction ($q_1 q_2 < 0$), we have $k = |q_1 q_2|/(4\pi\epsilon_0)$.

periodic solutions of the Kepler problem have negative total energies $E = -\langle K \rangle = (1/2)\langle U \rangle$.

The general solution of the Kepler problem involves solutions for the radial position $r(t)$ and angular position $\theta(t)$

$$\mu \ddot{r} = \frac{\ell^2}{\mu r^3} - \frac{k}{r^2} \equiv -V'(r) \quad \text{and} \quad \dot{\theta} = \frac{\ell}{\mu r^2}, \qquad (4.20)$$

whose orbits $r(\theta)$ are either bounded (periodic) or unbounded (see Fig. 4.2) in the effective potential

$$V(r) = -\frac{k}{r} + \frac{\ell^2}{2\mu r^2}. \qquad (4.21)$$

To obtain an analytic solution $r(\theta)$ for the Kepler problem (4.20), as expressed by the radial force equation (4.5), we use the normalized central potential $\overline{U}(s) = -s_0 s$, where $s_0 = \mu k/\ell^2$, and Eq. (4.5) becomes

$$s'' + s = s_0. \qquad (4.22)$$

The general solution to this equation is $s(\theta) = s_0 + \alpha \cos \theta$, where α is a constant and we have chosen the x-axis as the location where s has its minimum and maximum.

Next, the turning points for the Kepler problem are solutions of the quadratic equation

$$s^2 - 2 s_0 s - \epsilon = 0,$$

which can be written as $s_{1,2} = s_0 \pm \sqrt{s_0^2 + \epsilon}$:

$$s_1 = s_0(1 - e) \quad \text{and} \quad s_2 = s_0(1 + e),$$

where the eccentricity is defined as

$$e = \sqrt{1 + \epsilon/s_0^2} = \sqrt{1 + 2E\ell^2/\mu k^2}.$$

We now clearly see that $s(\theta) = s_0(1 + e \cos \theta)$ and the general solution to the Kepler Problem is

$$r(\theta) = \frac{r_0}{1 + e \cos \theta}, \qquad (4.23)$$

where $r_0 = 1/s_0 \equiv r(\pi/2)$ denotes the position of the minimum of the effective potential $V'(r_0) = 0$ (see Fig. 4.2). We note that motion is bounded (i.e., orbits are periodic) when $V_0 = -k/(2r_0) \leq E < 0$ ($0 \leq e < 1$), and the motion is unbounded (i.e., orbits are aperiodic) when $E \geq 0$ ($e \geq 1$).

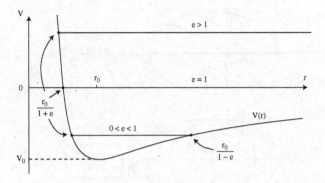

Fig. 4.2 Effective potential (4.21) for the Kepler problem.

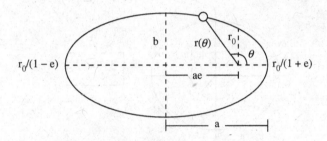

Fig. 4.3 Elliptical orbit for the Kepler problem.

4.3.1 *Bounded Keplerian Orbits*

4.3.1.1 *Kepler's First Law*

We first look at the bounded case ($\epsilon < 0$ or $e < 1$). Equation (4.23) generates an ellipse of semi-major axis

$$2\,a = \frac{r_0}{1+e} + \frac{r_0}{1-e} \;\;\rightarrow\;\; a = \frac{r_0}{1-e^2} = \frac{k}{2\,|E|} \qquad (4.24)$$

and semi-minor axis

$$b = a\sqrt{1-e^2} = \sqrt{\ell^2/(2\mu\,|E|)}. \qquad (4.25)$$

With these definitions, Eq. (4.23) may be written as

$$\left(\frac{x}{a} + e\right)^2 + \frac{y^2}{b^2} = 1, \qquad (4.26)$$

which yields Kepler's First Law: Planets move around the Sun along elliptical orbits.

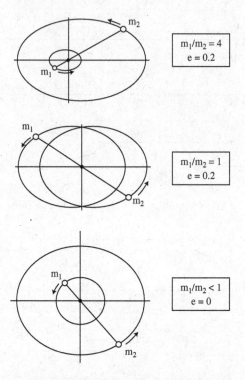

Fig. 4.4   Keplerian two-body orbits for various mass ratios and eccentricities.

When we plot the positions of the two objects (of mass m_1 and m_2, respectively) by using Kepler's first law (4.23), with the positions $\mathbf{r}_1$ and $\mathbf{r}_2$ determined by Eqs. (2.53), we obtain Fig. 4.4. It is interesting to note that by detecting the small wobble motion of a distant star (with mass m_1), it has been possible to discover extra-solar planets (with masses $m_2 < m_1$).

4.3.1.2   *Kepler's Second Law*

Using the conservation law of angular momentum (4.3), we find

$$dt = \frac{d\theta}{\dot{\theta}} = \frac{\mu}{\ell}\, r^2\, d\theta = \frac{2\mu}{\ell}\, dA(\theta),$$

where $dA(\theta) = (\int r\, dr)\, d\theta = \frac{1}{2}\,[r(\theta)]^2\, d\theta$ denotes an infinitesimal area swept by $d\theta$ at radius $r(\theta)$. When integrated, the relation

$$\Delta t = \frac{2\mu}{\ell}\, \Delta A, \tag{4.27}$$

yields Kepler's Second law: Equal areas ΔA are swept in equal times Δt since μ and ℓ are constants.

4.3.1.3 *Kepler's Third Law*

Using Kepler's Second Law (4.27), the orbital period T of a bound system is defined as

$$T = \int_0^{2\pi} \frac{d\theta}{\dot{\theta}} = \frac{\mu}{\ell} \int_0^{2\pi} r^2 \, d\theta = \frac{2\mu}{\ell} A = \frac{2\pi\mu}{\ell} a b$$

where $A = \pi a b$ denotes the area of an ellipse with semi-major axis a and semi-minor axis b. Next, using the expressions (4.24)-(4.25) for a and b, the orbital period becomes

$$T = \frac{2\pi\mu}{\ell} \cdot \frac{k}{2|E|} \cdot \sqrt{\frac{\ell^2}{2\mu|E|}} = 2\pi \sqrt{\frac{\mu k^2}{(2|E|)^3}}.$$

If we now substitute the expression for $a = k/2|E|$ and square both sides of this equation, we obtain Kepler's Third Law:

$$T^2 = \frac{(2\pi)^2\mu}{k} a^3. \tag{4.28}$$

In Newtonian gravitational theory, where $k/\mu = G(m_1 + m_2)$, Kepler's Third Law states that T^2/a^3 is a constant for all planets in the solar system, which is only an approximation that holds for $m_1 \gg m_2$ (true for all solar planets).

4.3.2 *Unbounded Keplerian Orbits*

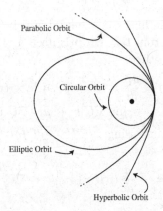

Fig. 4.5 Bounded and unbounded orbits for the Kepler problem.

We now look at the case where the total energy is positive or zero, i.e., $e \geq 1$ in Eq. (4.23). By defining $a = r_0/(e^2 - 1)$ and $b = r_0/\sqrt{e^2 - 1}$,

Eq. (4.23) yields the hyperbolic equation

$$\left(\frac{x}{a} - \text{e}\right)^2 - \frac{y^2}{b^2} = 1.$$

Furthermore, we may use a parametric form for the coordinates $x = a\,(\text{e} - \cosh\psi)$ and $y = b\sinh\psi$ and find that the hyperbolic asymptotes (at $\psi \to \pm\infty$) are located at $\pi \pm \Theta$, where $\Theta \equiv \arctan(\sqrt{\text{e}^2 - 1}) = \arccos(\text{e}^{-1})$.

When $\text{e} = 1$, Eq. (4.23) yields $r + x = r_0$, from which we recover the parabola $x = (r_0^2 - y^2)/2r_0$, with distance of closest approach reached at $x(0) = r_0/2$.

4.3.3 *Laplace-Runge-Lenz Vector**

Since the orientation of the unperturbed Keplerian ellipse is constant (i.e., it does not precess), it turns out there exists a third constant of the motion for the Kepler problem (in addition to energy and angular momentum); we note, however, that only two of these three invariants are independent.

4.3.3.1 *Kepler Problem*

Let us now investigate this additional constant of the motion for the Kepler problem. First, we consider the time derivative of the vector $\mathbf{p} \times \mathbf{L}$, where the linear momentum $\mathbf{p}$ and angular momentum $\mathbf{L}$ are

$$\mathbf{p} = \mu\left(\dot{r}\,\widehat{\mathbf{r}} + r\dot{\theta}\,\widehat{\theta}\right) \quad \text{and} \quad \mathbf{L} = \ell\,\widehat{\mathbf{z}} = \mu r^2\dot{\theta}\,\widehat{\mathbf{z}}.$$

The time derivative of the linear momentum is $\dot{\mathbf{p}} = -\nabla U(r) = -U'(r)\,\widehat{\mathbf{r}}$ while the angular momentum $\mathbf{L} = \mathbf{r} \times \mathbf{p}$ is itself a constant of the motion (in a central potential) so that

$$\frac{d}{dt}(\mathbf{p} \times \mathbf{L}) = \frac{d\mathbf{p}}{dt} \times \mathbf{L} = -\mu\,\nabla U \times (\mathbf{r} \times \dot{\mathbf{r}})$$

$$= -\mu\dot{\mathbf{r}}\cdot\nabla U\,\mathbf{r} + \mu\mathbf{r}\cdot\nabla U\,\dot{\mathbf{r}}.$$

By re-arranging terms (and using $\dot{\mathbf{r}}\cdot\nabla U = dU/dt$ for time-independent potentials), we find

$$\frac{d}{dt}(\mathbf{p} \times \mathbf{L}) = -\frac{d}{dt}(\mu U\,\mathbf{r}) + \mu\,(\mathbf{r}\cdot\nabla U + U)\,\dot{\mathbf{r}},$$

or

$$\frac{d\mathbf{A}}{dt} = \mu\,(\mathbf{r}\cdot\nabla U + U)\,\dot{\mathbf{r}}, \tag{4.29}$$

where $\mathbf{A}$ is the Laplace-Runge-Lenz (LRL) vector:

$$\mathbf{A} = \mathbf{p} \times \mathbf{L} + \mu U(r)\,\mathbf{r}. \tag{4.30}$$

We immediately note that the LRL vector (4.30) is a constant of the motion if the potential $U(r)$ satisfies the condition

$$\mathbf{r} \cdot \nabla U(r) + U(r) = \frac{d(r\,U)}{dr} = 0.$$

This condition is satisfied for the Kepler problem, with $U(r) = -k/r$, so that the LRL vector (4.30)

$$\mathbf{A} = \mathbf{p} \times \mathbf{L} - k\mu\hat{\mathbf{r}} = \left(\frac{\ell^2}{r} - k\mu\right)\hat{\mathbf{r}} - \ell\,\mu\dot{r}\,\hat{\theta}, \qquad (4.31)$$

is a constant of the motion.

Since the vector $\mathbf{A}$ is constant in both magnitude and direction, its constant magnitude

$$|\mathbf{A}|^2 = 2\mu\ell^2\left(\frac{p^2}{2\mu} + U\right) + k^2\mu^2 = k^2\mu^2\left(1 + \frac{2\ell^2 E}{\mu k^2}\right) = k^2\mu^2 e^2$$

is expressed in terms of $e(E, \ell)$. Next, we choose its direction to be along the x-axis ($\mathbf{A} = k\mu e\,\hat{x}$) and we can easily show that

$$\left(\frac{\ell^2}{r} - k\mu\right) = \mathbf{A} \cdot \hat{\mathbf{r}} \equiv (k\mu e)\cos\theta$$

leads to the Kepler solution (4.23), where $r_0 = \ell^2/k\mu$ and the orbit's eccentricity is $e = |\mathbf{A}|/k\mu$.

4.3.3.2 *Perturbed Kepler Problem*

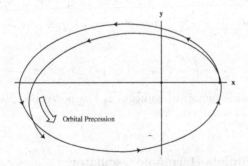

Fig. 4.6   Perturbed Kepler problem.

Note that if the Keplerian orbital motion is perturbed by the introduction of an additional potential term $\delta U(r)$, we find

$$\frac{d\mathbf{A}}{dt} = (\delta U + \mathbf{r} \cdot \nabla\delta U)\,\mathbf{p} \equiv \frac{d}{dr}(r\,\delta U)\,\mathbf{p},$$

where $\mathbf{A} = \mathbf{A}_0 + \mu \, \delta U \, \mathbf{r}$ is no longer a constant of the perturbed motion ($\mathbf{A}_0$ denotes the unperturbed LRL vector).

We now show that, under the perturbation potential $\delta U(r)$, the bounded (periodic) orbit precesses in θ. First, we obtain the cross product (to lowest order in δU)

$$\mathbf{A}_0 \times \frac{d\mathbf{A}}{dt} = (\delta U + \mathbf{r} \cdot \nabla \delta U) \, (p^2 + \mu \, U) \, \mathbf{L},$$

where $U = -k/r$ is the unperturbed Kepler potential. Next, using the expression for the unperturbed total energy

$$E = \frac{p^2}{2\mu} + U = -\frac{k}{2a},$$

we define the precession frequency

$$\omega_{\mathrm{p}}(\theta) = \widehat{\mathbf{z}} \cdot \frac{\mathbf{A}_0}{|\mathbf{A}_0|^2} \times \frac{d\mathbf{A}}{dt} = (\delta U + \mathbf{r} \cdot \nabla \delta U) \, \frac{\ell \, \mu}{(\mu k e)^2} \, (2 E - U)$$

$$= (\delta U + \mathbf{r} \cdot \nabla \delta U) \, \frac{\ell \mu k}{(\mu k e)^2} \left(\frac{1}{r} - \frac{1}{a} \right).$$

Hence, using $a = r_0/(1 - e^2)$, the precession frequency becomes

$$\omega_{\mathrm{p}}(\theta) = \ell^{-1} \left(1 + e^{-1} \cos\theta \right) \frac{d}{dr} \left(r \, \delta U \right), \qquad (4.32)$$

and the net precession shift $\delta\theta$ of the perturbed Keplerian orbit over one unperturbed period is

$$\delta\theta = \int_0^{2\pi} \omega_{\mathrm{p}}(\theta) \, \frac{d\theta}{\dot{\theta}} = \int_0^{2\pi} \left(\frac{1 + e^{-1} \cos\theta}{1 + e \cos\theta} \right) \left[r \, \frac{d}{dr} \left(\frac{r \, \delta U}{k} \right) \right]_{r = r(\theta)} d\theta.$$

For example, if $\delta U = -\epsilon/r^2$, then $r \, d(r\delta U/k)/dr = \epsilon/kr$ and the net precession shift is

$$\delta\theta = \frac{\epsilon}{k r_0} \int_0^{2\pi} \left(1 + e^{-1} \cos\theta \right) d\theta = 2\pi \, \frac{\epsilon}{k r_0}.$$

Figure 4.6 shows the numerical solution of the perturbed Kepler problem for the case where $\epsilon \simeq k r_0/16$.

4.4   Isotropic Simple Harmonic Oscillator

As a second example of a central potential with closed bounded orbits [see Eq. (4.19) with $n = 2$], we now investigate the case when the central potential is of the form

$$U(r) = \frac{k}{2} \, r^2 \quad \rightarrow \quad \overline{U}(s) = \frac{\mu k}{2 \ell^2 \, s^2}. \qquad (4.33)$$

The turning points for this problem are expressed as

$$r_1 = r_0 \left(\frac{1-e}{1+e}\right)^{\frac{1}{4}} = \frac{1}{s_1} \quad \text{and} \quad r_2 = r_0 \left(\frac{1+e}{1-e}\right)^{\frac{1}{4}} = \frac{1}{s_2},$$

where $r_0 = \sqrt{r_1 r_2} = (\ell^2/\mu k)^{1/4} = 1/s_0$ is the radial position at which the effective potential has a minimum, i.e., $V'(r_0) = 0$ and $V_0 = V(r_0) = k r_0^2$ and

$$e = \sqrt{1 - \left(\frac{k r_0^2}{E}\right)^2} = \sqrt{1 - \left(\frac{V_0}{E}\right)^2}.$$

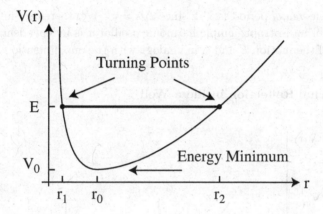

Fig. 4.7   Effective potential for the isotropic simple harmonic oscillator problem.

Here, we see from Fig. 4.7 that orbits are always bounded for $E > V_0$ (and thus $0 \le e \le 1$). Next, using the change of coordinate $q = s^2$ in Eq. (4.12), we obtain

$$\theta = \frac{-1}{2} \int_{q_2}^{q} \frac{dq}{\sqrt{\varepsilon q - q_0^2 - q^2}}, \tag{4.34}$$

where $q_2 = (1 + e)\varepsilon/2$ and $q_0 = s_0^2$. We now substitute $q(\varphi) = (1 + e \cos \varphi)\varepsilon/2$ in Eq. (4.34) to obtain

$$\theta(q) = \frac{1}{2} \arccos\left[\frac{1}{e}\left(\frac{2q}{\varepsilon} - 1\right)\right],$$

and we easily verify that $\Delta\theta = \pi$ and bounded orbits are closed. This equation can now be inverted to give

$$r(\theta) = \frac{r_0 (1 - e^2)^{1/4}}{\sqrt{1 + e \cos 2\theta}}, \tag{4.35}$$

which describes the ellipse $x^2/b^2 + y^2/a^2 = 1$, with semi-major axis $a = r_2$ and semi-minor axis $b = r_1$. Note that this solution $x^2/b^2 + y^2/a^2 = 1$ may be obtained from the Cartesian representation for the Lagrangian $L = \frac{1}{2}\mu(\dot{x}^2 + \dot{y}^2) - \frac{1}{2}k(x^2 + y^2)$, which yields the solutions $x(t) = b\cos\omega t$ and $y(t) = a\sin\omega t$, where the constants a and b are determined from the conservation laws $E = \frac{1}{2}\mu\omega^2(a^2 + b^2)$ and $\ell = \mu\omega\, a\, b$.

Lastly, the area of the ellipse is $A = \pi ab = \pi r_0^2$ while the *physical* period is

$$T(E, \ell) = \int_0^{2\pi} \frac{d\theta}{\dot{\theta}} = \frac{2\mu A}{\ell} = 2\pi\sqrt{\frac{\mu}{k}};$$

note that the *radial* period is $T/2$ since $\Delta\theta = \pi$. We, therefore, find that the period of an isotropic simple harmonic oscillator is independent of the constants of the motion E and ℓ, in analogy with the one-dimensional case.

4.5   Internal Reflection inside a Well

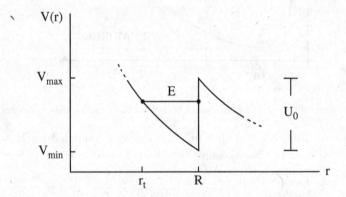

Fig. 4.8   Effective potential for the internal hard sphere.

As a last example of bounded motion in a central-force potential, we consider the constant central potential

$$U(r) = \begin{cases} -U_0 & (r < R) \\ 0 & (r > R) \end{cases}$$

where U_0 is a constant and R denotes the radius of a sphere. The effective potential $V(r) = \ell^2/(2\mu r^2) + U(r)$ associated with this potential is shown in

Fig. 4.8. Orbits are unbounded when $E > V_{\max} = \ell^2/(2\mu R^2)$. For energy values

$$V_{\min} = \frac{\ell^2}{2\mu R^2} - U_0 < E < V_{\max} = \frac{\ell^2}{2\mu R^2},$$

on the other hand, Fig. 4.8 shows that bounded motion is possible, with turning points

$$r_1 = r_t = \sqrt{\frac{\ell^2}{2\mu(E+U_0)}} \quad \text{and} \quad r_2 = R.$$

When $E = V_{\min}$, the left turning point reaches its maximum value $r_t = R$ while it reaches its minimum value $r_t/R = (1 + U_0/E)^{-\frac{1}{2}} < 1$ when $E = V_{\max}$.

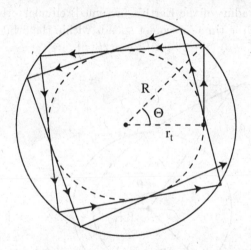

Fig. 4.9   Internal reflections inside a hard sphere.

Assuming that the particle starts at $r = r_t$ at $\theta = 0$, the particle orbit is found by integration by quadrature as

$$\theta(s) = \int_s^{s_t} \frac{d\sigma}{\sqrt{s_t^2 - \sigma^2}},$$

where $s_t = 1/r_t$, which is easily integrated to yield

$$\theta(s) = \arccos\left(\frac{s}{s_t}\right) \quad \rightarrow \quad r(\theta) = r_t \sec\theta \quad (\text{for } \theta \leq \Theta),$$

where the maximum angle Θ defines the angle at which the particle hits the turning point R, i.e., $r(\Theta) = R$ and

$$\Theta = \arccos\left(\sqrt{\frac{\ell^2}{2\mu R^2 (E + U_0)}}\right).$$

Subsequent motion of the particle involves an infinite sequence of *internal reflections* as shown in Fig. 4.9. The case where $E > \ell^2/2\mu R^2$ involves a single turning point and is discussed in Sec. 5.6.2.

4.6 Problems

1. Consider a comet moving in a parabolic orbit in the plane of the Earth's orbit. If the distance of closest approach of the comet to the sun is $\beta\, r_E$, where r_E is the radius of the Earth's (assumed) circular orbit and where $\beta < 1$, show that the time the comet spends within the orbit of the Earth is given by $\sqrt{2\,(1-\beta)}\,(1+2\,\beta) \times 1\ \text{year}/(3\,\pi)$.

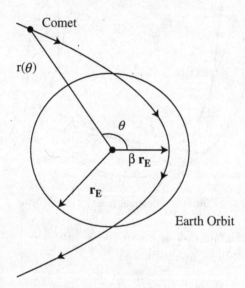

Fig. 4.10   Problem 1.

2. Find the effective potential $V(r) = U(r) + \ell^2/(2\mu r^2)$ that allows a particle to move in a spiral orbit given by $r = k\,\theta^n$, where k is a constant and n is a positive integer.

3. A satellite (of mass m) is orbiting Earth at an altitude h (above ground) along a circular path with total energy E and angular momentum ℓ. During its orbit, the satellite's propulsion system gives an instantaneous boost $v \to v + \Delta v$ to the satellite (with Δv tangent to the orbit).

(a) Find the new Keplerian orbit if $\Delta v > 0$.

(b) Find the minimum boost value $\Delta v_{\min}$ necessary for the satellite to escape Earth's gravity (i.e., the satellite's after-boost orbit is a parabola).

4. Show that the radial "fall-in" time of a particle of mass m in a potential $U = -k/r$ starting from radius R without angular momentum is

$$\pi \sqrt{\frac{m R^3}{8 k}}$$

5. Consider the perturbed Kepler problem in which a particle of mass m, energy $E < 0$, and angular momentum ℓ is moving in the central-force potential

$$U(r) = -\frac{k}{r} + \frac{\alpha}{r^2},$$

where the perturbation potential α/r^2 is considered small in the sense that the dimensionless parameter $\epsilon = 2m\alpha/\ell^2 \ll 1$ is small.

(a) Show that the energy equation for this problem can be written using $s = 1/r$ as

$$E = \frac{\ell^2}{2m} \left[(s')^2 + \gamma^2 s^2 - 2 s_0 s \right],$$

where $s_0 = mk/\ell^2$ and $\gamma^2 = 1 + \epsilon$.

(b) Show that the turning points are

$$s_1 = \frac{s_0}{\gamma^2} (1 - e) \quad \text{and} \quad s_2 = \frac{s_0}{\gamma^2} (1 + e),$$

where $e = \sqrt{1 + 2 \gamma^2 \ell^2 E/mk^2}$.

(c) By solving the integral

$$\theta(s) = -\int_{s_2}^{s} \frac{d\sigma}{\sqrt{(2mE/\ell^2) + 2 s_0\sigma - \gamma^2 \sigma^2}},$$

where $\theta(s_2) = 0$, show that

$$r(\theta) = \frac{\gamma^2 r_0}{1 + e \cos(\gamma\theta)},$$

where $r_0 = 1/s_0$. Hence, the ellipse precesses with an angular step $\Delta\theta = 2\pi/\gamma$.

6. Consider a particle of mass m moving in the potential $U(\mathbf{r}) = -k/r - \mathbf{r} \cdot \mathbf{F}$, where $\mathbf{F}$ is a constant force vector. Show that, while the angular momentum $\mathbf{L}$ is no longer conserved, the quantity

$$\mathbf{p} \times \mathbf{L} \cdot \mathbf{F} - \frac{mk}{r} \mathbf{r} \cdot \mathbf{F} + \frac{m}{2} |\mathbf{r} \times \mathbf{F}|^2$$

is a constant of the motion.

7. A Keplerian elliptical orbit, described by the relation $r(\theta) = r_0/(1 + e \cos\theta)$, undergoes a precession motion when perturbed by the perturbation potential $\delta U(r)$, with precession frequency (4.32). Show that, if $\delta U(r) = -\alpha/r^3$ (where α is a constant), the net precession shift $\delta\theta$ of the Keplerian orbit over one unperturbed period is

$$\delta\theta = \int_0^{2\pi} \omega_p(\theta) \frac{d\theta}{\dot\theta} = 6\pi \frac{\alpha}{kr_0^2}.$$

8. In Kepler's work, angles are referred to as *anomalies*. In Fig. 4.11, an ellipse (with eccentricity $e < 1$) of semi-major axis a and semi-minor axis b is inscribed by a circle of radius a.

(a) Show that the orbit of the planet (at point P in Fig. 4.11) is described in terms of the *eccentric anomaly* ψ as

$$r(\psi) = a(1 - e \cos\psi),$$

and the *true* anomaly θ is defined in terms of ψ as

$$\cos\theta(\psi) = \left(\frac{\cos\psi - e}{1 - e \cos\psi} \right).$$

Note that by using the eccentric anomaly angle ψ, we find $a \cos\psi = ae + r \cos\theta$ from which we obtain $\cos\psi = (e + \cos\theta)/(1 + e \cos\theta)$ or $\cos\theta = (\cos\psi - e)/(1 - e \cos\psi)$. By substituting this last expression into Kepler's First Law (4.23), we obtain $r(\psi) = a(1 - e \cos\psi)$.

(b) Show that the time from perihelion ($\psi = 0$) is given by Kepler's Equation:

$$t(\psi) = \frac{\tau}{2\pi} (\psi - e \sin\psi), \tag{4.36}$$

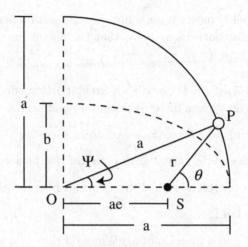

Fig. 4.11   Problem 8.

where $\tau \equiv 2\pi \sqrt{\mu a^3/k}$ denotes the orbital period (i.e., 1 year for Earth).
(c) If the Earth's orbit is divided in two by the *latus rectum* (i.e., the vertical line drawn through the Sun), show that the times spent in the inner and outer halves (in fractions of a year) are

$$t_{\text{inner}} = \frac{1}{\pi}\left[\cos^{-1}e \, - \, e\sqrt{1-e^2}\right],$$

$$t_{\text{outer}} = \frac{1}{\pi}\left[(\pi - \cos^{-1}e) \, + \, e\sqrt{1-e^2}\right],$$

and that the difference between the times is

$$\Delta t \equiv t_{\text{outer}} - t_{\text{inner}} = \frac{2}{\pi}\left[\sin^{-1}e + e\sqrt{1-e^2}\right].$$

(d) Using $t(\psi)$ and $r(\psi)$, show that the average orbital radius is

$$\bar{r} = a\left(1 + \frac{e^2}{2}\right).$$

9. A particle of unit mass moves from infinity along a straight line that, if continued would allow it to pass a distance $b\sqrt{2}$ from a point P. If the particle is attracted toward point P with a force varying as k/r^5, and if the angular momentum about the point P is $\sqrt{k}/b$, show that the trajectory is given by

$$r = b\coth(\theta/\sqrt{2}).$$

10. An Earth satellite moves in an elliptical orbit with a period τ, eccentricity e, and semi-major axis a. Show that the maximum radial velocity of the satellite is $2\pi\, ae/(\tau\,\sqrt{1-e^2})$.

11. (a) Show that if a particle describes a circular orbit under the influence of an attractive central force directed toward a point on the circle,

$$r(\theta) \;=\; 2\,R\,|\cos\theta| \;\equiv\; \sqrt{2}\,R\,\sqrt{1+\cos 2\theta},$$

then the force varies as the inverse-fifth power of the distance.

(b) Show that for the orbit described the total energy of the particle is zero.

(c) Find the period of the motion.

(d) Find $\dot{x}$, $\dot{y}$, and v as a function of angle around the circle and show that all three quantities are inifinite as the particle goes the center of force.

12*. At perigee of an elliptic gravitational orbit, a particle experiences an impulse in the radial direction, sending the particle into another elliptic orbit. Determine the new semi-major axis, eccentricity, and orientation in terms of the old.

13. Show that for elliptical motion in a gravitational field the radial speed can be written as

$$\dot{r} \;=\; \omega\,\frac{a}{r}\,\sqrt{a^2 e^2 \;-\; (r-a)^2}.$$

By replacing the radial coordinate r with the eccentric anomaly angle ψ, show that the resulting differential equation can be integrated immediately to give Kepler's Equation (4.36).

14. Discuss the possible types of orbit for a particle moving under the central potential $k/2r^2$. (a) For the repulsive case $(k > 0)$, show that the orbit equation is $r(\theta) = b\sec n(\theta - \theta_0)$, where n, b, and θ_0 are constants.

(b) For the attractive case $(k < 0)$, the nature of the orbit depends on the sign of $\ell^2 + mk$ and E. Find the orbit equation for each possible type.

15. The *Hohmann* transfer orbit (H) represents the passage from a circular orbit at radius r_A to another circular orbit at radius $r_B > r_A$ (see inset in Fig. 4.12).

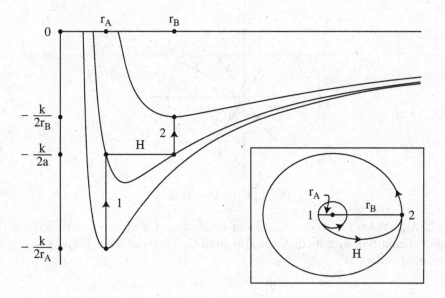

Fig. 4.12 Problem 15.

The transfer orbit requires two boosts at points 1 and 2, with energy changes

$$\Delta E_1 = \frac{k}{2}\left(\frac{1}{r_A} - \frac{1}{a}\right) \quad \text{and} \quad \Delta E_2 = \frac{k}{2}\left(\frac{1}{a} - \frac{1}{r_B}\right),$$

where $a = (r_A + r_B)/2$ is the semi-major axis for the Hohmann (H) transfer orbit. For each boost ($j = 1, 2$), compute the velocity change Δv_j.

16. Consider a binary-star system composed of two stars of masses m_1 and $m_2 < m_1$ undergoing circular motion about their common center of mass (see Fig. 4.4).

(a) Use Kepler's Third Law to show that the masses m_1 and m_2 are expressed in terms of the orbital radii r_1 and r_2 according to the relations

$$m_1 = \frac{\omega^2 a^2}{G} r_2 \quad \text{and} \quad m_2 = \frac{\omega^2 a^2}{G} r_1,$$

where $\omega \equiv 2\pi/T$ is the orbital frequency of the two stars and $a = r_1 + r_2$ is the constant separation between the two stars.

(b) Compute the masses of Sirius A and Sirius B (in units of solar mass) if their separation is 20 AU ($1\,\text{AU} = 1.5 \times 10^{11}\,\text{m}$), with $r_B/r_A = 2.3$, and the orbital period is $T = 50$ years.

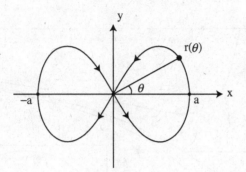

Fig. 4.13   Problem 17.

17. A particle of mass m and angular momentum ℓ is observed to undergo periodic motion with its distance $r(\theta)$ from the center of force given by the relation

$$r^2(\theta) \;=\; a^2 \, \cos(2\,\theta),$$

which describes a *lemniscate* of amplitude a (see Fig. 4.13).

(a) Show that the potential energy $U(r)$ leading to this periodic motion is given by the attractive potential

$$U(r) \;=\; -\,\frac{\ell^2}{2m}\,\frac{a^4}{r^6}.$$

(b) Solve the differential equation for $\theta(t)$ given by $\dot{\theta} = \ell/mr^2$ for $\theta < \pi/4$ subject to the initial condition $\theta = 0$ at time $t = 0$.

18. Show that for a particle moving along an elliptical orbit in an inverse-square-law potential, the eccentricity of the orbit can be written as

$$e \;=\; \frac{\sqrt{n}-1}{\sqrt{n}+1},$$

where $n = \dot{\varphi}_{\mathrm{max}}/\dot{\varphi}_{\mathrm{min}}$ is the ratio of the maximum angular velocity to the minimum angular velocity.

Chapter 5

Collisions and Scattering Theory

In Chapter 4, we investigated two types of orbits (bounded and unbounded) for two-particle systems evolving under the influence of a central potential. In the present Chapter, we focus our attention on unbounded orbits within the context of *elastic* collision theory (i.e., a collision for which energy and momentum are conserved). In this context, a collision between two interacting particles involves a three-step process (see Fig. 5.1): Step I – two particles are initially infinitely far apart (in which case, the energy of each particle is assumed to be strictly kinetic); Step II – as the two particles approach each other, their interacting potential (repulsive or attractive) causes them to reach a distance of closest approach (where the interaction force is strongest); and Step III – the two particles then move progressively farther apart (eventually reaching a point where the energy of each particle is once again strictly kinetic).

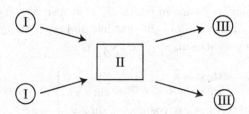

Fig. 5.1   Collision kinematics ($I \rightarrow III$) and dynamics (II).

These three steps form the foundations of Collision *Kinematics* and Collision *Dynamics*. The topic of Collision Kinematics, which describes the collision in terms of the conservation laws of momentum and energy, deals with Steps I and III; here, the incoming particles define the initial state of the two-particle system while the outgoing particles define the final state. The topic of Collision Dynamics, on the other hand, deals with Step II, in which the particular nature of the interaction is taken into account.

5.1   Two-Particle Collisions in the LAB Frame

Consider the collision of two particles (labeled 1 and 2) of masses m_1 and m_2, respectively. Let us denote the velocities of particles 1 and 2 *before* the collision as $\mathbf{u}_1$ and $\mathbf{u}_2$, respectively, while the velocities *after* the collision are denoted $\mathbf{v}_1$ and $\mathbf{v}_2$. Furthermore, the particle momenta before and after the collision are denoted $\mathbf{p}$ and $\mathbf{q}$, respectively.

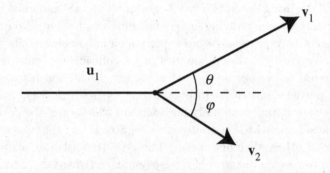

Fig. 5.2   Collision kinematics in the LAB frame.

To simplify the analysis, we define the laboratory (LAB) frame to correspond to the reference frame in which m_2 is at rest (i.e., $\mathbf{u}_2 = 0$); in this collision scenario, m_1 is the *projectile* particle and m_2 is the *target* particle. We now write the velocities $\mathbf{u}_1$, $\mathbf{v}_1$, and $\mathbf{v}_2$ as

$$\left. \begin{array}{l} \mathbf{u}_1 = u\,\widehat{\mathbf{x}} \\ \mathbf{v}_1 = v_1\,(\cos\theta\,\widehat{\mathbf{x}} + \sin\theta\,\widehat{\mathbf{y}}) \\ \mathbf{v}_2 = v_2\,(\cos\varphi\,\widehat{\mathbf{x}} - \sin\varphi\,\widehat{\mathbf{y}}) \end{array} \right\} , \tag{5.1}$$

where the *deflection* angle θ of the projectile particle and the *recoil* angle φ of the target particle are defined in Fig. 5.2. The conservation laws of momentum and energy

$$m_1\mathbf{u}_1 = m_1\mathbf{v}_1 + m_2\mathbf{v}_2 \quad \text{and} \quad \frac{m_1}{2}\,u^2 = \frac{m_1}{2}\,|\mathbf{v}_1|^2 + \frac{m_2}{2}\,|\mathbf{v}_2|^2$$

can be written in terms of the mass ratio $\alpha = m_1/m_2$ of the projectile mass to the target mass as

$$\alpha \left(u - v_1 \cos \theta\right) = v_2 \cos \varphi, \tag{5.2}$$

$$\alpha \, v_1 \sin \theta = v_2 \sin \varphi, \tag{5.3}$$

$$\alpha \left(u^2 - v_1^2\right) = v_2^2. \tag{5.4}$$

Since the **three** equations (5.2)-(5.4) are expressed in terms of **four** unknown quantities $(v_1, \theta; v_2, \varphi)$, for given incident velocity u and mass ratio α, we must choose **one** post-collision coordinate as an independent variable to get a closed solution for the three remaining post-collision variables.

Here, we choose the recoil angle φ of the target particle, and proceed with finding expressions for $v_1(u, \varphi; \alpha)$, $v_2(u, \varphi; \alpha)$ and $\theta(u, \varphi; \alpha)$; other choices lead to similar formulas (see problems at the end of the Chapter). First, adding the square of the momentum components (5.2) and (5.3), we obtain

$$\alpha^2 v_1^2 = \alpha^2 u^2 - 2 \alpha \, u v_2 \cos \varphi + v_2^2. \tag{5.5}$$

Next, using the energy equation (5.4), we find

$$\alpha^2 v_1^2 = \alpha \left(\alpha u^2 - v_2^2\right) = \alpha^2 u^2 - \alpha v_2^2, \tag{5.6}$$

so that these two equations combine to give

$$v_2(u, \varphi; \alpha) = 2 \left(\frac{\alpha}{1 + \alpha}\right) u \cos \varphi. \tag{5.7}$$

Once $v_2(u, \varphi; \alpha)$ is known and after substituting Eq. (5.7) into Eq. (5.6), we find

$$v_1(u, \varphi; \alpha) = u \sqrt{1 - 4 \frac{\mu}{M} \cos^2 \varphi}, \tag{5.8}$$

where $\mu/M = \alpha/(1 + \alpha)^2$ is the ratio of the reduced mass μ and the total mass M.

Lastly, we take the ratio of the momentum components (5.2) and (5.3) in order to eliminate the unknown v_1 and find

$$\tan \theta = \frac{v_2 \sin \varphi}{\alpha u - v_2 \cos \varphi}.$$

If we substitute Eq. (5.7), we easily obtain

$$\tan \theta = \frac{2 \sin \varphi \cos \varphi}{1 + \alpha - 2 \cos^2 \varphi},$$

or

$$\theta(\varphi; \alpha) = \arctan \left(\frac{\sin 2\varphi}{\alpha - \cos 2\varphi}\right). \tag{5.9}$$

In the limit $\alpha = 1$ (i.e., a collision involving identical particles), we find $v_2 = u \cos \varphi$ and $v_1 = u \sin \varphi$ from Eqs. (5.7) and (5.8), respectively, and

$$\tan \theta = \cot \varphi \quad \to \quad \varphi = \frac{\pi}{2} - \theta,$$

from Eq. (5.9). Hence, the angular sum $\theta + \varphi$ for like-particle collisions is always 90° (for $\varphi \neq 0$).

We summarize by stating that, after the collision, the momenta $\mathbf{q}_1$ and $\mathbf{q}_2$ in the LAB frame (where m_2 is initially at rest) are functions of the initial momentum $p = m_1 u$ and the angles θ and φ:

$$\mathbf{q}_1 = p \left[1 - \frac{4\alpha}{(1+\alpha)^2} \cos^2 \varphi \right]^{1/2} (\cos\theta\, \hat{\mathbf{x}} + \sin\theta\, \hat{\mathbf{y}}), \qquad (5.10)$$

$$\mathbf{q}_2 = \frac{2p \cos\varphi}{1+\alpha} (\cos\varphi\, \hat{\mathbf{x}} - \sin\varphi\, \hat{\mathbf{y}}). \qquad (5.11)$$

We note that Eqs. (5.10)-(5.11) satisfy the law of conservation of (kinetic) energy in addition to the law of conservation of momentum.

5.2   Two-Particle Collisions in the CM Frame

In the center-of-mass (CM) frame (see Sec. 2.6), the elastic collision between particles 1 and 2 is described quite simply; the CM velocities and momenta are, henceforth, denoted with a prime. Before the collision, the momenta of particles 1 and 2 are equal in magnitude but with opposite directions (see Fig. 5.3)

$$\mathbf{p}_1' = \mu u\, \hat{\mathbf{x}} = -\, \mathbf{p}_2',$$

where μ is the reduced mass of the two-particle system. After the collision, conservation of energy-momentum dictates that

$$\mathbf{q}_1' = \mu u\, (\cos\Theta\, \hat{\mathbf{x}} + \sin\Theta\, \hat{\mathbf{y}}) = -\, \mathbf{q}_2', \qquad (5.12)$$

where Θ is the scattering angle in the CM frame and $\mu u = p/(1+\alpha)$. Thus the particle velocities after the collision in the CM frame are

$$\mathbf{v}_1' = \frac{\mathbf{q}_1'}{m_1} = \frac{u}{1+\alpha} (\cos\Theta\, \hat{\mathbf{x}} + \sin\Theta\, \hat{\mathbf{y}}) \quad \text{and} \quad \mathbf{v}_2' = \frac{\mathbf{q}_2'}{m_2} = -\, \alpha\, \mathbf{v}_1'.$$

It is quite clear, thus, that the initial and final kinematic states lie on the same circle in CM momentum space and the single variable defining the outgoing two-particle state is represented by the CM scattering angle Θ.

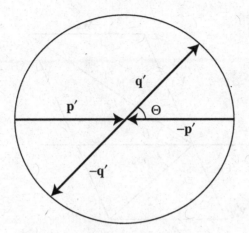

Fig. 5.3 Collision kinematics in the CM frame.

5.3 . Connection between the CM and LAB Frames

We now establish the connection between the momenta (5.10)-(5.11) in the LAB frame and the momenta (5.12) in the CM frame (see Fig. 5.4). First, we denote the velocity of the CM as

$$\mathbf{w} = \frac{m_1 \mathbf{u}_1 + m_2 \mathbf{u}_2}{m_1 + m_2} = \frac{\alpha u}{1 + \alpha} \, \widehat{\mathsf{x}},$$

so that $w = |\mathbf{w}| = \alpha u/(1 + \alpha)$ and $|\mathbf{v}_2'| = w = \alpha |\mathbf{v}_1'|$.

The connection between $\mathbf{v}_1'$ and $\mathbf{v}_1$ is expressed as

$$\mathbf{v}_1 = \mathbf{v}_1' + \mathbf{w} \quad \rightarrow \quad \begin{cases} v_1 \cos\theta = w \, (1 + \alpha^{-1} \cos\Theta) \\[2mm] v_1 \sin\theta = w \, \alpha^{-1} \sin\Theta \end{cases}$$

so that

$$\tan\theta = \frac{\sin\Theta}{\alpha + \cos\Theta}, \tag{5.13}$$

and

$$v_1 = \frac{u}{(1+\alpha)} \sqrt{1 + \alpha^2 + 2\alpha \cos\Theta}. \tag{5.14}$$

Likewise, the connection between $\mathbf{v}_2'$ and $\mathbf{v}_2$ is expressed as

$$\mathbf{v}_2 = \mathbf{v}_2' + \mathbf{w} \quad \rightarrow \quad \begin{cases} v_2 \cos\varphi = w \, (1 - \cos\Theta) \\[2mm] v_2 \sin\varphi = w \sin\Theta \end{cases}$$

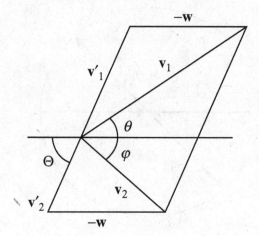

Fig. 5.4   CM and LAB collision geometries.

so that $\tan \varphi = \sin \Theta / (1 - \cos \Theta) = \cot \Theta / 2$, which yields

$$\varphi = \frac{1}{2} (\pi - \Theta), \qquad (5.15)$$

and

$$v_2 = \frac{2\alpha\, u}{(1 + \alpha)}\, \sin \frac{\Theta}{2}. \qquad (5.16)$$

5.4   Scattering Cross Sections

In the previous Section, we investigated the connection between the initial and final kinematic states of an elastic collision described by Steps I and III, respectively, introduced earlier (see Fig. 5.1). Here, the initial kinematic state is described in terms of the speed u of the projectile particle in the Laboratory frame (assuming that the target particle is at rest), while the final kinematic state is described in terms of the velocity coordinates for the deflected projectile particle (v_1, θ) and the recoiled target particle (v_2, φ).

In the present Section, we shall investigate Step II, namely, how the distance of closest approach influences the deflection angles (θ, φ) in the LAB frame and the deflection angle Θ in the CM frame.

5.4.1 Definitions

First, we consider for simplicity the case of a projectile particle of mass m being deflected by a repulsive central-force potential $U(r) > 0$ whose center is at rest at the origin (i.e., $\alpha = 0$). As the projectile particle approaches from the right (at $r = \infty$ and $\theta = 0$) moving with speed u, it is progressively deflected until it reaches a minimum radius ρ at $\theta = \chi$ after which the projectile particle moves away from the repulsion center until it reaches $r = \infty$ at a deflection angle $\theta = \Theta$ and again moving with speed u.

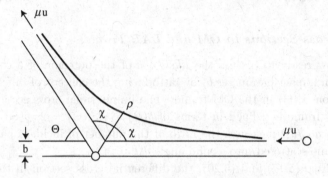

Fig. 5.5 Scattering geometry: the incoming particle is entering the interaction region from the right and is leaving on an asymptote after having been deflected by an angle Θ.

Figure 5.5 shows that the scattering process is symmetric about the line of closest approach (i.e., $2\chi = \pi - \Theta$, where Θ is the CM deflection angle). The angle of closest approach

$$\chi = \frac{1}{2}(\pi - \Theta) \qquad (5.17)$$

is a function of the distance of closest approach ρ, the total energy E, and the angular momentum ℓ. The distance ρ is, of course, a turning point ($\dot{r} = 0$) and is the only positive root of the energy equation

$$E = U(\rho) + \frac{\ell^2}{2m\rho^2}, \qquad (5.18)$$

where $E = mu^2/2$ is the total initial energy of the projectile particle.

The path of the projectile particle in Fig. 5.5 is labeled by the *impact parameter* b (the distance of closest approach in the non-interacting case: $U = 0$). A simple calculation (using $\mathbf{r} = -ut\widehat{\mathbf{x}} + b\widehat{\mathbf{y}}$ and $\mathbf{u} = -u\widehat{\mathbf{x}}$) shows that the angular momentum $\mathbf{L} = \ell\widehat{\mathbf{z}}$ is

$$\ell = \widehat{\mathbf{z}} \cdot (m\mathbf{r} \times \mathbf{u}) = mub = \sqrt{2mE}\,b. \qquad (5.19)$$

It is, thus, quite clear that ρ is a function of $E \equiv \ell^2/(2m\,b^2)$, m, and b. The angle χ in Fig. 5.5 is defined as

$$\chi = \int_{\rho}^{\infty} \frac{(\ell/r^2)\,dr}{\sqrt{2m\,[E - U(r)] - (\ell^2/r^2)}}$$

$$\equiv \int_{0}^{b/\rho} \frac{dx}{\sqrt{1 - x^2 - b^2\,\overline{U}(x/b)}}. \tag{5.20}$$

Once an expression $\Theta(b) \equiv \pi - 2\,\chi(b)$ is obtained from Eq. (5.20), we may invert it to obtain $b(\Theta)$.

5.4.2   *Cross Sections in CM and LAB Frames*

We are now ready to discuss the *likelihood* of the outcome of a collision (for a given impact parameter b) by introducing the concept of differential cross section $\sigma'(\Theta)$ in the CM frame. The infinitesimal cross section $d\sigma'$ in the CM frame is defined in terms of $b(\Theta)$ as $d\sigma'(\Theta) = 2\pi\,b(\Theta)\,db(\Theta)$. Physically, $d\sigma'/d\Omega$ measures the ratio of the number of incident particles per unit time scattered into a solid angle $d\Omega$.

Using Eqs. (5.17) and (5.20), the differential cross section in the CM frame is defined as

$$\sigma'(\Theta) = \frac{d\sigma'}{2\pi \sin\Theta\,d\Theta} = \frac{b(\Theta)}{\sin\Theta} \left| \frac{db(\Theta)}{d\Theta} \right|, \tag{5.21}$$

where we note that the quantity $db/d\Theta$ is often negative and, thus, we must take its absolute value to ensure that $\sigma'(\Theta)$ is positive. The total cross section is defined as

$$\sigma_T = 2\pi \int_{0}^{\pi} \sigma'(\Theta)\,\sin\Theta\,d\Theta. \tag{5.22}$$

The differential cross section can also be written in the LAB frame in terms of the deflection angle θ as

$$\sigma(\theta) = \frac{d\sigma}{2\pi \sin\theta\,d\theta} = \frac{b(\theta)}{\sin\theta} \left| \frac{db(\theta)}{d\theta} \right|. \tag{5.23}$$

Since the infinitesimal cross section $d\sigma = d\sigma'$ is the same in both frames (i.e., the likelyhood of a collision should not depend on the choice of a frame of reference), we find

$$\sigma(\theta)\,\sin\theta\,d\theta = \sigma'(\Theta)\,\sin\Theta\,d\Theta,$$

from which we obtain

$$\sigma(\theta) = \sigma'(\Theta)\,\frac{\sin\Theta}{\sin\theta}\,\frac{d\Theta}{d\theta}, \tag{5.24}$$

or

$$\sigma'(\Theta) = \sigma(\theta) \frac{\sin\theta}{\sin\Theta} \frac{d\theta}{d\Theta}. \tag{5.25}$$

Eq. (5.24) yields an expression for the differential cross section in the LAB frame $\sigma(\theta)$ once the differential cross section in the CM frame $\sigma'(\Theta)$ and an explicit formula for $\Theta(\theta)$ are known. Equation (5.25) represents the inverse transformation $\sigma(\theta) \rightarrow \sigma'(\Theta)$. We point out that, whereas the CM differential cross section $\sigma'(\Theta)$ is naturally associated with theoretical calculations, the LAB differential cross section $\sigma(\theta)$ is naturally associated with experimental measurements. Hence, the transformation (5.24) is used to translate a theoretical prediction into an observable experimental cross section, while the transformation (5.25) is used to translate experimental measurements into a format suitable for theoretical analysis.

We note that these transformations rely on finding relations between the LAB deflection angle θ and the CM deflection angle Θ given by Eq. (5.13), which can be converted into

$$\sin(\Theta - \theta) = \alpha \sin\theta. \tag{5.26}$$

For example, using these relations, we now show how to obtain an expression for Eq. (5.24) by using Eqs. (5.13) and (5.26). First, we use Eq. (5.26) to obtain

$$\frac{d\Theta}{d\theta} = \frac{\alpha\cos\theta + \cos(\Theta - \theta)}{\cos(\Theta - \theta)}, \tag{5.27}$$

where

$$\cos(\Theta - \theta) = \sqrt{1 - \alpha^2 \sin^2\theta}\ .$$

Next, using Eqs. (5.13) and (5.26), we show that

$$\frac{\sin\Theta}{\sin\theta} = \frac{\alpha + \cos\Theta}{\cos\theta} = \frac{\alpha + [\cos(\Theta - \theta)\cos\theta - \sin(\Theta - \theta)\sin\theta]}{\cos\theta}$$

$$= \frac{\alpha(1 - \sin^2\theta) + \cos(\Theta - \theta)\cos\theta}{\cos\theta}$$

$$= \alpha\cos\theta + \sqrt{1 - \alpha^2\sin^2\theta}\ . \tag{5.28}$$

Thus by combining Eqs. (5.27) and (5.28), we find

$$\frac{\sin\Theta}{\sin\theta}\frac{d\Theta}{d\theta} = \frac{[\alpha\cos\theta + \sqrt{1 - \alpha^2\sin^2\theta}]^2}{\sqrt{1 - \alpha^2\sin^2\theta}}$$

$$= 2\alpha\cos\theta + \frac{1 + \alpha^2\cos 2\theta}{\sqrt{1 - \alpha^2\sin^2\theta}}, \tag{5.29}$$

which is valid for $\alpha < 1$. Lastly, noting from Eq. (5.26) that the CM deflection angle is defined as

$$\Theta(\theta) = \theta + \arcsin(\alpha \sin\theta),$$

the transformation $\sigma'(\Theta) \to \sigma(\theta)$ is now complete. Similar manipulations yield the transformation $\sigma(\theta) \to \sigma'(\Theta)$. We note that the LAB-frame cross section $\sigma(\theta)$ are generally difficult to obtain for arbitrary mass ratio $\alpha = m_1/m_2$.

5.5 Rutherford Scattering

As an explicit example of the scattering formalism developed in this Chapter, we investigate the scattering of a charged particle of mass m_1 and charge q_1 by another charged particle of mass $m_2 \gg m_1$ and charge q_2 such that $q_1 q_2 > 0$ and $\mu \simeq m_1$. This situation leads to the two particles experiencing a repulsive central force with potential

$$U(r) = \frac{k}{r},$$

where $k = q_1 q_2 / (4\pi\varepsilon_0) > 0$.

The turning-point equation in this case is

$$E = E \frac{b^2}{\rho^2} + \frac{k}{\rho},$$

whose solution is the distance of closest approach

$$\rho = r_0 + \sqrt{r_0^2 + b^2} = b\left(\epsilon + \sqrt{1 + \epsilon^2}\right), \qquad (5.30)$$

where $2r_0 = k/E$ is the distance of closest approach for a *head-on* collision (for which the impact parameter b is zero) and $\epsilon = r_0/b$. Note, here, that the second solution $r_0 - \sqrt{r_0^2 + b^2}$ to the turning-point equation is negative and, therefore, is not allowed. The problem of the electrostatic repulsive interaction between a positively-charged alpha particle (i.e., the nucleus of a helium atom) and positively-charged nucleus of a gold atom was first studied by Rutherford (Ernest Rutherford, 1871-1937) and the scattering cross section for this problem is known as the Rutherford cross section.

The angle χ at which the distance of closest approach is reached is calculated from Eq. (5.20) as

$$\chi = \int_0^{b/\rho} \frac{dx}{\sqrt{1 - x^2 - 2\epsilon x}} = \int_0^{b/\rho} \frac{dx}{\sqrt{(1 + \epsilon^2) - (x + \epsilon)^2}}, \qquad (5.31)$$

where
$$\frac{b}{\rho} = \frac{1}{\epsilon + \sqrt{1+\epsilon^2}} = -\epsilon + \sqrt{1+\epsilon^2}.$$

Making use of the trigonometric substitution $x = -\epsilon + \sqrt{1+\epsilon^2}\,\cos\psi$, we find that
$$\chi = \arccos\left(\frac{\epsilon}{\sqrt{1+\epsilon^2}}\right) \quad \rightarrow \quad \epsilon = \cot\chi,$$

which becomes
$$b = r_0 \tan\chi. \tag{5.32}$$

Using the relation (5.17), we now find
$$b(\Theta) = r_0 \cot\frac{\Theta}{2}. \tag{5.33}$$

Since $db(\Theta)/d\Theta = -(r_0/2)\,\csc^2(\Theta/2)$, we take its absolute value and obtain from Eq. (5.21) the CM Rutherford cross section
$$\sigma'(\Theta) = \frac{b(\Theta)}{\sin\Theta}\left|\frac{db(\Theta)}{d\Theta}\right| = \frac{r_0^2}{4\,\sin^4(\Theta/2)},$$

or
$$\sigma'(\Theta) = \left(\frac{k}{4E\,\sin^2(\Theta/2)}\right)^2. \tag{5.34}$$

Note that the Rutherford scattering cross section (5.34) does not depend on the sign of k and is thus valid for both repulsive and attractive interactions.

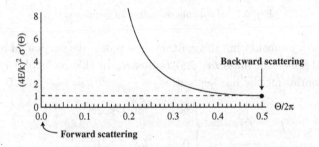

Fig. 5.6   Rutherford scattering cross-section.

Figure 5.6 shows that the Rutherford scattering cross section (5.34) becomes very large in the forward direction $\Theta \to 0$ (where $\sigma' \to \Theta^{-4}$) while the differential cross section behaves as $\sigma' \to (k/4E)^2$ as $\Theta \to \pi$ (i.e., backward scattering). Note that the forward-scattering divergence of the Rutherford formula (5.34) can be eliminated by a slight modification of the potential $U(r)$ (see Eq. (5.46) and problems at the end of the Chapter).

5.6   Hard-Sphere and Soft-Sphere Scattering

Explicit calculations of differential cross sections tend to be very complex for general central potentials. In the present Section, we consider two simple central potentials associated with a uniform central potential $U(r) \neq 0$ confined to a spherical region $(r < R)$.

5.6.1   *Hard-Sphere Scattering*

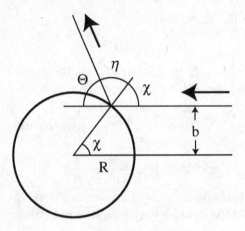

Fig. 5.7   Hard-sphere scattering geometry.

We begin by considering the collision of a point-like particle of mass m_1 with a hard sphere of mass m_2 and radius R. In this particular case, the central potential for the hard sphere is

$$
U(r) = \begin{cases} \infty & (\text{for } r < R) \\ 0 & (\text{for } r > R) \end{cases}
$$

and the collision is shown in Fig. 5.7, where we see that the impact parameter

$$b = R \sin \chi, \tag{5.35}$$

depends simply on the angle of incidence χ. The angle of reflection η is different from the angle of incidence χ for the case of arbitrary mass ratio $\alpha = m_1/m_2$.

To show this, we decompose the velocities in terms of components perpendicular and tangential to the surface of the sphere at the point of impact, i.e., we respectively find

$$\alpha\, u\, \cos\chi = v_2 - \alpha\, v_1\, \cos\eta$$
$$\alpha\, u\, \sin\chi = \alpha\, v_1\, \sin\eta.$$

From these expressions we obtain

$$\tan\eta = \frac{\alpha\, u\, \sin\chi}{v_2 - \alpha\, u\, \cos\chi}.$$

From Fig. 5.7, we also find the deflection angle $\theta = \pi - (\chi + \eta)$ and the recoil angle $\varphi = \chi$. Hence, substituting Eq. (5.7), we find

$$\tan\eta = \left(\frac{1+\alpha}{1-\alpha}\right)\tan\chi. \tag{5.36}$$

We, therefore, easily see that $\eta = \chi$ (the standard form of the Law of Reflection) only if $\alpha = 0$ (i.e., the target particle is infinitely massive).

In the CM frame (with $\alpha = 0$), the collision is symmetric with a deflection angle $\chi = \frac{1}{2}(\pi - \Theta)$, so that

$$b = R\sin\chi = R\cos\frac{\Theta}{2}.$$

The scattering cross section in the CM frame is

$$\sigma'(\Theta) = \frac{b(\Theta)}{\sin\Theta}\left|\frac{db(\Theta)}{d\Theta}\right| = \frac{R\cos(\Theta/2)}{\sin\Theta}\cdot\left|-\frac{R}{2}\sin(\Theta/2)\right| = \frac{R^2}{4}, \tag{5.37}$$

and the total cross section is

$$\sigma_T = 2\pi\int_0^\pi \sigma'(\Theta)\sin\Theta\, d\Theta = \pi R^2, \tag{5.38}$$

i.e., the total cross section for the problem of hard-sphere collision is equal to the effective area of the sphere.

The scattering cross section in the LAB frame can also be obtained for the case $\alpha < 1$ using Eqs. (5.24) and (5.29) as

$$\sigma(\theta) = \frac{R^2}{4}\left(2\alpha\,\cos\theta + \frac{1+\alpha^2\,\cos2\theta}{\sqrt{1-\alpha^2\sin^2\theta}}\right), \tag{5.39}$$

for $\alpha = m_1/m_2 < 1$. The integration of this formula yields the total cross section

$$\sigma_T = 2\pi\int_0^\pi \sigma(\theta)\sin\theta\, d\theta \equiv \pi R^2,$$

where $\theta_{max} = \pi$ for $\alpha < 1$ (see problem 10).

5.6.2 *Soft-Sphere Scattering*

We now consider the scattering of a particle subjected to the attractive potential considered in Sec. 4.5:

$$U(r) = \begin{cases} -U_0 & \text{(for } r < R) \\ 0 & \text{for } r > R \end{cases} \tag{5.40}$$

where the constant U_0 denotes the depth of the attractive potential well and the condition $E > \ell^2/2\mu R^2$ involves a single turning point. We denote β the angle at which the incoming particle enters the *soft-sphere* potential (see Fig. 5.8), and thus the impact parameter b of the incoming particle is $b = R \sin\beta$.

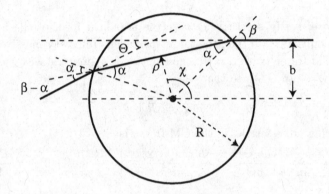

Fig. 5.8 Soft-sphere scattering geometry.

The particle enters the soft-sphere potential region $(r < R)$ and reaches a distance of closest approach ρ, defined from the turning-point condition

$$E = -U_0 + E\,\frac{b^2}{\rho^2} \quad \rightarrow \quad \rho = \frac{b}{\sqrt{1 + U_0/E}} = \frac{R}{n}\sin\beta,$$

where $n = \sqrt{1 + U_0/E} > 1$ denotes the *index of refraction* of the soft-sphere potential region. From Fig. 5.8, we note that an optical analogy helps us determine that, through Snell's law, we find

$$\sin\beta = n \sin\left(\beta - \frac{\Theta}{2}\right), \tag{5.41}$$

where the *transmission* angle α is given in terms of the *incident* angle β and the CM scattering angle $-\Theta$ is defined as $\Theta = 2(\beta - \alpha)$.

The distance of closest approach is reached at an angle χ is determined as

$$\chi = \beta + \int_{\rho}^{R} \frac{b\,dr}{r\sqrt{n^2r^2 - b^2}}$$

$$= \beta + \arccos\left(\frac{b}{nR}\right) - \underbrace{\arccos\left(\frac{b}{n\rho}\right)}_{= \, 0}$$

$$= \beta + \arccos\left(\frac{b}{nR}\right) = \frac{1}{2}\left(\pi + \Theta\right), \tag{5.42}$$

and, thus, the impact parameter $b(\Theta)$ can be expressed as $b(\Theta) = nR\sin(\beta(b) - \Theta/2)$ or, if we solve for $\beta(b)$ explicitly, as

$$b(\Theta) = \frac{nR\sin(\Theta/2)}{\sqrt{1 + n^2 - 2n\cos(\Theta/2)}}. \tag{5.43}$$

Its derivative with respect to Θ yields

$$\frac{db}{d\Theta} = \frac{nR}{2}\frac{[n\cos(\Theta/2) - 1]\,[n - \cos(\Theta/2)]}{[1 + n^2 - 2n\cos(\Theta/2)]^{3/2}},$$

and the scattering cross section (5.21) in the CM frame is

$$\sigma'(\Theta) = \frac{n^2R^2}{4}\frac{|[n\cos(\Theta/2) - 1]\,[n - \cos(\Theta/2)]|}{\cos(\Theta/2)\,[1 + n^2 - 2n\cos(\Theta/2)]^2}. \tag{5.44}$$

Note that, on the one hand, when $\beta = 0$, we find $\chi = \pi/2$ and $\Theta_{\min} = 0$ while, on the other hand, when $\beta = \pi/2$, we find $b = R$ and

$$1 = n\sin\left(\frac{\pi}{2} - \frac{\Theta_{\max}}{2}\right) = n\cos(\Theta_{\max}/2),$$

which yields the maximum angle

$$\Theta_{\max} = 2\arccos\left(n^{-1}\right).$$

Moreover, when $\Theta = \Theta_{\max}$, we find that $db/d\Theta$ vanishes and, therefore, the differential cross section vanishes $\sigma'(\Theta_{\max}) = 0$, while at $\Theta = 0$, we find $\sigma'(0) = [n/(n-1)]^2\,(R^2/4)$.

Figure 5.9 shows the soft-sphere scattering cross section $\overline{\sigma}(\Theta) \equiv (4/R^2)\,\sigma'(\Theta)$ as a function of Θ for four cases: $n = (1.1, 1.15)$ in the *soft-sphere limit* $(n \to 1)$ and $n = (10, 20, 50, 1000)$ in the *hard-sphere limit* $(n \to \infty)$. We clearly see the strong forward-scattering behavior as $n \to 1$ (or $U_0 \to 0$) in the soft-sphere limit and the hard-sphere limit $\overline{\sigma} \to 1$ as

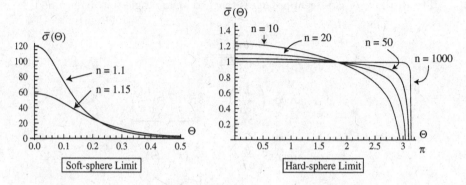

Fig. 5.9 Soft-sphere scattering cross-section in the soft-sphere limit $(n \to 1)$ and the hard-sphere limit $(n \gg 1)$; here, note that $\bar{\sigma}(0) = n^2/(n-1)^2$.

$n \to \infty$. We note that, using the substitution $x = n \cos \Theta/2$, the total scattering cross section associated with Eq. (5.44)

$$\sigma_T = 2\pi \int_0^{\Theta_{max}} \sigma'(\Theta) \, \sin \Theta \, d\Theta$$

$$= 2\pi R^2 \int_1^n \frac{(x-1)\,(n^2-x)\,dx}{(1 + n^2 - 2x)^2} = \pi R^2 \qquad (5.45)$$

is independent of the index of refraction n and equals the hard-sphere total cross section (5.38).

The soft-sphere potential (5.40) plays an important role in providing a simple tool to investigate neutron scattering experiments with heavy nuclei. Note that we can construct a realistic model of the nuclear potential $U(r)$ for scattering experiments with positively-charged particles by combining the attractive soft-sphere potential (5.40) with the Coulomb potential:

$$U(r) = \begin{cases} -U_0 & (r < R) \\ \\ k/r & (r > R) \end{cases} \qquad (5.46)$$

where R represents the nuclear radius.

5.7 Elastic Scattering by a Hard Surface

We now generalize the hard-sphere scattering problem by considering scattering by a smooth hard surface of revolution[1] $\rho(z)$ with maximal radial

---

[1] Adapted from J. L. Brun and A. F. Pacheco, Euro. J. Phys. **26**, 747 (2005).

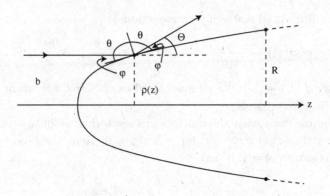

Fig. 5.10   Scattering by a hard surface $\rho(z)$.

extent R (see Fig. 5.10). Here, a particle of mass m, initially traveling along the z-axis with velocity u with an impact parameter $b < R$, collides with the hard surface and is scattered with deflection angle Θ. The particle hits the surface at a distance $b = \rho(z)$ from its axis of symmetry and the angle of incidence $\theta = \pi/2 - \varphi$ (measured from the normal to the surface) is defined in terms of the complementary angle φ, where $\cos\varphi = [1+(\rho')^2]^{-1/2}$. Since the deflection angle Θ is defined in terms of φ as $\Theta = \pi - 2\theta = 2\varphi$, we find

$$\tan\varphi = \rho'(z) = \tan\frac{\Theta}{2}. \tag{5.47}$$

By using the identity $b(\Theta) = \rho(z)$, we can solve for $z(\Theta)$ [or $\Theta(z)$], and we can thus calculate the differential cross-section (5.21).

First, we use the identity

$$\frac{db}{d\Theta} = \rho'\,\frac{dz}{d\Theta} = \rho'\left(\frac{d\Theta}{dz}\right)^{-1},$$

where, by inverting Eq. (5.47), we obtain $\Theta(z) = 2\arctan(\rho')$, which yields

$$\frac{d\Theta}{dz} = \frac{2\rho''}{[1+(\rho')^2]},$$

and, hence,

$$\left|\frac{db}{d\Theta}\right| = \frac{\rho'}{2\,|\rho''|}\,[1+(\rho')^2].$$

Lastly, using the relation

$$\sin\Theta = 2\cos\varphi\,\sin\varphi = \frac{2\rho'}{[1+(\rho')^2]},$$

we find the differential scattering cross-section

$$\sigma(\Theta(z)) = \frac{\rho}{4\,|\rho''|}\left[1+(\rho')^2\right]^2 \equiv \left(\frac{\rho}{4\,\kappa}\right)\sec\frac{\Theta}{2}, \qquad (5.48)$$

where $\kappa \equiv |\rho''|/[1+(\rho')^2]^{3/2}$ denotes the Frenet-Serret curvature of the curve $\rho(z)$ in the (ρ, z)-plane.

For example, we revisit the hard-sphere scattering problem studied in Sec. 5.6.1, with $\rho(z) = \sqrt{R^2 - z^2}$ for $-R \leq z \leq 0$. Here, the Frenet-Serret curvature is simply $\kappa = 1/R$ and

$$z = -\rho\tan\frac{\Theta}{2} \;\rightarrow\; \rho = R\cos\frac{\Theta}{2},$$

so that the differential cross-section (5.48) yields the standard hard-sphere result (5.37):

$$\sigma(\Theta(z)) = \left(\frac{R^2}{4}\cos\frac{\Theta}{2}\right)\sec\frac{\Theta}{2} = \frac{R^2}{4}.$$

Note that it is possible to invert the relation $\rho(z) \rightarrow \sigma(\Theta(z))$, given by Eq. (5.48), to obtain the shape of a surface from its scattering data $\sigma(\Theta) \rightarrow \rho(z(\Theta))$.

5.8   Problems

1. (a) A particle of mass m_1 traveling in a straight line at velocity v_1 has a head-on elastic collision with a particle of mass m_2 traveling at velocity $-v_2$ (along the same line but in the opposite direction). Show that after the collision, the masses m_1 and m_2 are assumed to travel at velocities $-v_1'$ and v_2' defined as

$$v_1' = \frac{(1-\alpha)\,v_1 + 2\,v_2}{(1+\alpha)} \quad\text{and}\quad v_2' = \frac{2\alpha\,v_1 - (1-\alpha)\,v_2}{(1+\alpha)},$$

where $\alpha \equiv m_1/m_2$.

(b) Consider the transfer of momentum from a particle of mass M, initially traveling at velocity u, to another of the same mass M, mediated by a third particle of mass $m < M$. The particles are arranged in a straight line, with the lighter particle placed in the middle (initially at rest), so that the collision process begins when the heavier particle collides head-on with the lighter particle, which then collides with the third particle (also initially at rest). Find the fraction of the initial momentum that is finally transferred to the last particle as a function of $\alpha \equiv m/M < 1$.

2. (a) Using Eq. (5.1) and the conservation laws of energy and momentum, solve for $v_1(u, \theta; \beta)$, where $\beta = m_2/m_1$.

(b) Discuss the number of physical solutions for $v_1(u, \theta; \beta)$ for $\beta < 1$ and $\beta > 1$.

(c) For $\beta < 1$, show that physical solutions for $v_1(u, \theta; \beta)$ exist for $\theta < \arcsin(\beta) = \theta_{max}$.

3. Show that the momentum transfer $\Delta \mathbf{p}'_1 = \mathbf{q}'_1 - \mathbf{p}'_1$ of the projectile particle in the CM frame has a magnitude

$$|\Delta \mathbf{p}'_1| = 2\,\mu u \, \sin \frac{\Theta}{2},$$

where μ, u, and Θ are the reduced mass, initial projectile LAB speed, and CM scattering angle, respectively.

4. Show that the differential cross section $\sigma'(\Theta)$ for the elastic scattering of a particle of mass m from the repulsive central-force potential $U(r) = k/r^2$ with a fixed force-center at $r = 0$ (or an infinitely massive target particle) is

$$\sigma'(\Theta) = \frac{2\pi^2 k}{m\,u^2} \frac{(\pi - \Theta)}{[\Theta\,(2\pi - \Theta)]^2 \sin \Theta},$$

where u is the speed of the incoming projectile particle at $r = \infty$.

Hint :   Show that $b(\Theta) = \dfrac{r_0\,(\pi - \Theta)}{\sqrt{2\pi\,\Theta - \Theta^2}},$ *where* $r_0^2 = \dfrac{2\,k}{m\,u^2}.$

5. By using the relations $\tan \theta = \sin \Theta / (\alpha + \cos \Theta)$ and/or $\sin(\Theta - \theta) = \alpha \sin \theta$, where $\alpha = m_1/m_2$, show that the relation between the differential cross section in the CM frame, $\sigma'(\Theta)$, and the differential cross section in the LAB frame, $\sigma(\theta)$, is

$$\sigma'(\Theta) = \sigma(\theta) \cdot \frac{1 + \alpha \cos \Theta}{(1 + 2\,\alpha \cos \Theta + \alpha^2)^{3/2}}.$$

6. Consider the scattering of a particle of mass m by the localized repulsive central potential

$$U(r) = \begin{cases} -kr^2/2 & (r \le R) \\ 0 & (r > R) \end{cases}$$

where the radius R denotes the range of the interaction.

(a) Show that for a particle of energy $E > 0$ moving towards the center of attraction with impact parameter $b = R \sin \beta$, the distance of closest approach ρ for this problem is

$$\rho = \sqrt{\frac{E}{k}} \, (e - 1), \quad \text{where} \quad e = \sqrt{1 + \frac{2 \, k b^2}{E}}$$

(b) Show that the angle χ at closest approach is

$$\chi = \beta + \int_{\rho}^{R} \frac{(b/r^2) \, dr}{\sqrt{1 - b^2/r^2 + k r^2/2E}}$$

$$= \beta + \frac{1}{2} \arccos \left(\frac{2 \sin^2 \beta - 1}{e} \right)$$

(c) Using the relation $\chi = \frac{1}{2} (\pi + \Theta)$ between χ and the CM scattering angle Θ, show that

$$e = \frac{\cos 2\beta}{\cos(2\beta - \Theta)} > 1.$$

7. Consider the scattering of a particle of mass m by the potential

$$U(r) = \frac{k}{r} + \frac{\beta}{r^2}.$$

(a) Show that the distance of closest approach is

$$\rho = r_0 + \sqrt{r_0^2 + b^2 + \beta/E},$$

where $r_0 = k/2E$ and $b = \ell/\sqrt{2m E}$ is the impact parameter.

(b) Show that the angle of closest approach χ satisfies the transcendental equation

$$b = \frac{r_0}{\alpha} \tan(\alpha \chi),$$

where $\alpha \equiv \sqrt{1 + \beta/(E \, b^2)}$. (We can easily recover the Rutherford expression when $\beta = 0$ and $\alpha = 1$.)

8. Consider the scattering of a particle of mass m by the modified Rutherford potential

$$U(r) = \begin{cases} k \, (1/r - 1/R) & (r \leq R) \\ 0 & (r > R) \end{cases}$$

(a) Show that the distance of closest approach is

$$\rho = \frac{b}{\alpha}\left(\frac{r_0}{b\alpha} + \sqrt{1 + \frac{r_0^2}{b^2\alpha^2}}\right),$$

where $r_0 = k/2E$, $b = \ell/\sqrt{2mE}$ is the impact parameter, and $\alpha = \sqrt{1 + k/ER}$.

(b) Show that the angle of closest approach χ is expressed as

$$\chi = \beta + \arccos\left(\frac{x + \sin\beta}{\sqrt{1 + x^2 + 2x\,\sin\beta}}\right),$$

where $x \equiv r_0/b$ and $\beta \equiv \arcsin(b/R)$.

(c) Show that the impact parameter satisfies the transcendental equation

$$b = r_0\left[\cot(\Theta/2)\,\cos\beta - \sin\beta\right],$$

which can be solved explicitly as

$$b = \frac{R}{\sqrt{1 + \mu^2\,\tan^2\Theta/2}},$$

where $\mu = 1 + R/r_0$. Verify that the standard Rutherford equation (5.33) is recovered in the limit $R \to \infty$ (or $\beta \to 0$).

(d) Using the relation $b(\Theta)$ from Part (c), show that the differential cross-section for this modified Rutherford potential is

$$\sigma(\Theta) = \frac{R^2}{4}\,\frac{(1+\lambda)}{(1 + \lambda\,\sin^2\Theta/2)^2},$$

where $\lambda \equiv \mu^2 - 1$.

(e) From the differential cross-section calculated in Part (d), show that the total cross-section is

$$\sigma_T = \pi R^2 \int_0^\pi \frac{(1+\lambda)\,\sin\Theta\,d\Theta}{(1 + \lambda\,\sin^2\Theta/2)^2} \doteq 2\pi R^2.$$

9. Consider elastic scattering by a hard ellipsoid $\rho(z) = \rho_0\sqrt{1 - (z/z_0)^2}$ $(-z_0 \le z \le 0)$, where $\rho_0 = z_0\sqrt{1 - e^2} \le z_0$ and $0 \le e < 1$ denotes the eccentricity of the ellipse in the (ρ, z)-plane.

(a) Show that the differential scattering cross-section is expressed as

$$\sigma(\Theta) = \frac{\rho_0^2\,(1 - e^2)}{4\,(1 - e^2\,\cos^2\frac{\Theta}{2})}.$$

(b) Show that the total cross-section σ_T is

$$\sigma_T = 2\pi \int_0^\pi \sigma(\Theta) \sin \Theta \, d\Theta = \pi \rho_0^2 \left[\left(\frac{1}{e^2} - 1 \right) \ln \left(\frac{1}{1 - e^2} \right) \right].$$

(c) Show that we recover the hard-sphere result $\sigma_T = \pi \rho_0^2$ in the limit $e \to 0$.

10. Show that the LAB-frame scattering cross-section (5.39) yields the same total cross-section $\sigma_T = \pi R^2$ as in the CM frame.

11. Consider the scattering problem associated with a repulsive soft-sphere potential, where $-U_0$ is replaced with U_0 in Eq. (5.40). By replacing $n = (1 + U_0/E)^{\frac{1}{2}}$ with $n = (1 - U_0/E)^{-\frac{1}{2}}$, show that Eq. (5.43) is replaced with $b(\Theta) = n^{-1} R \sin(\beta(b) + \Theta/2)$, or

$$b(\Theta) = \frac{R \sin(\Theta/2)}{\sqrt{1 + n^2 - 2n \cos(\Theta/2)}},$$

while Snell's law (5.41) is replaced with

$$\sin \left(\beta + \frac{\Theta}{2} \right) = n \sin \beta.$$

12. Show that the elastic scattering by hard surface $\rho(z) = 2\sqrt{Rz}$, where R is a constant, yields the Rutherford formula

$$\sigma = R^2 / \sin^4 \frac{\Theta}{2},$$

where $z = R \cot^2 \frac{\Theta}{2}$.

Chapter 6

Motion in a Non-Inertial Frame

A reference frame is said to be an *inertial* frame if the motion of particles in that frame is subject only to physical forces (e.g., forces that are derivable from a physical potential U such that $m\ddot{\mathbf{x}} = -\nabla U$). The Principle of Galilean Relativity (Sec. 2.5.3) states that the laws of physics are the same in all inertial frames and that all reference frames moving at constant velocity with respect to an inertial frame are also inertial frames. Hence, physical accelerations are identical in all inertial frames.

In contrast, a reference frame is said to be *non-inertial* if the motion of particles in that frame of reference violates the Principle of Galilean Relativity. Such non-inertial frames include all rotating frames and accelerated reference frames.

6.1 Time Derivatives in Rotating Frames

To investigate the relationship between inertial and non-inertial frames, we consider the time derivative of an arbitrary vector $\mathbf{A}(t)$ in two reference frames. The first reference frame is called the *fixed* (inertial) frame and is expressed in terms of the Cartesian coordinates $\mathbf{r}' = (x', y', z')$. The second reference frame is called the *rotating* (non-inertial) frame and is expressed in terms of the Cartesian coordinates $\mathbf{r} = (x, y, z)$. In Fig. 6.1, the rotating frame shares the same origin as the fixed frame (we remove this condition later) and the rotation angular velocity $\boldsymbol{\omega}$ of the rotating frame (with respect to the rotating frame) has components $(\omega_x, \omega_y, \omega_z)$.

Since observations can also be made in a rotating frame of reference, we decompose the vector $\mathbf{A}$ in terms of components A_i in the rotating frame (with unit vectors $\hat{\mathbf{x}}^i$). Thus, $\mathbf{A} = A_i\,\hat{\mathbf{x}}^i$ (using the summation rule) and

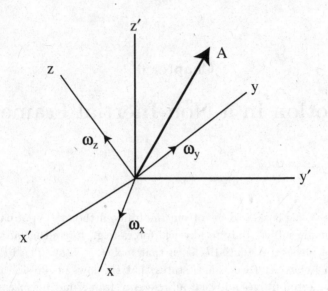

Fig. 6.1 Rotating and fixed frames.

the time derivative of $\mathbf{A}$ as observed in the fixed frame is

$$\frac{d\mathbf{A}}{dt} = \frac{dA_i}{dt}\,\widehat{\mathbf{x}}^i + A_i\,\frac{d\widehat{\mathbf{x}}^i}{dt}. \tag{6.1}$$

The interpretation of the first term is that of the time derivative of $\mathbf{A}$ as observed in the rotating frame (where the unit vectors $\widehat{\mathbf{x}}^i$ are constant) while the second-term involves the time-dependence of the relation between the fixed and rotating frames. By construction the vector $d\widehat{\mathbf{x}}^i/dt$ is simply expressed in terms of the angular velocity $\boldsymbol{\omega}$ of the rotating frame as

$$\frac{d\widehat{\mathbf{x}}^i}{dt} = \boldsymbol{\omega} \times \widehat{\mathbf{x}}^i. \tag{6.2}$$

Hence, the second term in Eq. (6.1) becomes

$$A_i \frac{d\widehat{\mathbf{x}}^i}{dt} = \boldsymbol{\omega} \times \mathbf{A}. \tag{6.3}$$

The time derivative of an arbitrary rotating-frame vector $\mathbf{A}$ in a fixed frame is, therefore, expressed as

$$\left(\frac{d\mathbf{A}}{dt}\right)_f = \left(\frac{d\mathbf{A}}{dt}\right)_r + \boldsymbol{\omega} \times \mathbf{A}, \tag{6.4}$$

where $(d/dt)_f$ denotes the time derivative as observed in the fixed (f) frame while $(d/dt)_r$ denotes the time derivative as observed in the rotating (r)

frame. An important application of this formula relates to the time derivative of the rotation angular velocity $\boldsymbol{\omega}$ itself. One can easily see that

$$\left(\frac{d\boldsymbol{\omega}}{dt}\right)_f = \dot{\boldsymbol{\omega}} = \left(\frac{d\boldsymbol{\omega}}{dt}\right)_r,$$

since the second term in Eq. (6.4) vanishes for $\mathbf{A} = \boldsymbol{\omega}$; the time derivative of $\boldsymbol{\omega}$ is, therefore, the same in both frames of reference and is denoted $\dot{\boldsymbol{\omega}}$ in what follows.

6.2 Accelerations in Rotating Frames

We now consider the general case of a rotating frame and fixed frame being related by translation and rotation. The position of a point P according to the fixed frame of reference is labeled $\mathbf{r}'$, while the position of the same point according to the rotating frame of reference is labeled $\mathbf{r}$, and

$$\mathbf{r}' = \mathbf{R} + \mathbf{r}, \tag{6.5}$$

where $\mathbf{R}$ denotes the position of the origin of the rotating frame (e.g., the center of mass) according to the fixed frame. Since the velocity of the point P involves the rate of change of position, we must now be careful in defining which time-derivative operator, $(d/dt)_f$ or $(d/dt)_r$, is used.

The velocities of point P as observed in the fixed and rotating frames are defined as

$$\mathbf{v}_f = \left(\frac{d\mathbf{r}'}{dt}\right)_f \quad \text{and} \quad \mathbf{v}_r = \left(\frac{d\mathbf{r}}{dt}\right)_r, \tag{6.6}$$

respectively. Using Eq. (6.4), the relation between the fixed-frame and rotating-frame velocities is expressed as

$$\mathbf{v}_f = \dot{\mathbf{R}} + \left(\frac{d\mathbf{r}}{dt}\right)_f = \dot{\mathbf{R}} + \mathbf{v}_r + \boldsymbol{\omega} \times \mathbf{r}, \tag{6.7}$$

where $\dot{\mathbf{R}}$ denotes the translation velocity of the rotating-frame origin (as observed in the fixed frame).

Using Eq. (6.7), we are now in a position to evaluate expressions for the acceleration of point P as observed in the fixed and rotating frames of reference, which are defined as

$$\mathbf{a}_f = \left(\frac{d\mathbf{v}_f}{dt}\right)_f \quad \text{and} \quad \mathbf{a}_r = \left(\frac{d\mathbf{v}_r}{dt}\right)_r, \tag{6.8}$$

respectively. Hence, using Eq. (6.7), we find

$$\mathbf{a}_f = \ddot{\mathbf{R}} + \left(\frac{d\mathbf{v}_r}{dt}\right)_f + \left(\frac{d\boldsymbol{\omega}}{dt}\right)_f \times \mathbf{r} + \boldsymbol{\omega} \times \left(\frac{d\mathbf{r}}{dt}\right)_f$$

$$= \ddot{\mathbf{R}} + (\mathbf{a}_r + \boldsymbol{\omega} \times \mathbf{v}_r) + \dot{\boldsymbol{\omega}} \times \mathbf{r} + \boldsymbol{\omega} \times (\mathbf{v}_r + \boldsymbol{\omega} \times \mathbf{r}),$$

or

$$\mathbf{a}_f = \ddot{\mathbf{R}} + \mathbf{a}_r + 2\boldsymbol{\omega} \times \mathbf{v}_r + \dot{\boldsymbol{\omega}} \times \mathbf{r} + \boldsymbol{\omega} \times (\boldsymbol{\omega} \times \mathbf{r}), \qquad (6.9)$$

where $\ddot{\mathbf{R}}$ denotes the translational acceleration of the rotating-frame origin (as observed in the fixed frame of reference). We can now write an expression for the acceleration of point P as observed in the rotating frame as

$$\mathbf{a}_r = \left(\mathbf{a}_f - \ddot{\mathbf{R}}\right) - \boldsymbol{\omega} \times (\boldsymbol{\omega} \times \mathbf{r}) - 2\boldsymbol{\omega} \times \mathbf{v}_r - \dot{\boldsymbol{\omega}} \times \mathbf{r}, \qquad (6.10)$$

which represents the sum of the net *inertial* acceleration $(\mathbf{a}_f - \ddot{\mathbf{R}})$, the centrifugal acceleration $-\boldsymbol{\omega} \times (\boldsymbol{\omega} \times \mathbf{r})$ and the *Coriolis* acceleration $-2\boldsymbol{\omega} \times \mathbf{v}_r$ (see Fig. 6.2) and an angular acceleration term $-\dot{\boldsymbol{\omega}} \times \mathbf{r}$ that depends explicitly on the time dependence of the rotation angular velocity $\boldsymbol{\omega}$. The centrifugal acceleration $\mathbf{a}_{\mathrm{Cf}} = -\boldsymbol{\omega} \times (\boldsymbol{\omega} \times \mathbf{r}) = \omega^2 \mathbf{r} - (\boldsymbol{\omega} \cdot \mathbf{r})\boldsymbol{\omega}$ (which is directed outwardly from the rotation axis) represents a familiar *non-inertial* effect in physics.

A less familiar *non-inertial* effect is the Coriolis acceleration $\mathbf{a}_{\mathrm{Co}} \equiv -2\boldsymbol{\omega} \times \dot{\mathbf{r}}$ discovered in 1831 by Gaspard Gustave de Coriolis (1792-1843). Figure 6.2 shows that an object *falling* inwardly (toward Earth) also experiences an *eastward* acceleration. It is also quite clear that, since the Coriolis acceleration does not change the kinetic energy of a particle (i.e., $\dot{\mathbf{r}} \cdot \mathbf{a}_{\mathrm{Co}} \equiv 0$), it only changes the direction of the particle's motion (if $\dot{\mathbf{r}}$ is not directed along $\boldsymbol{\omega}$).

6.3   Lagrangian Formulation of Non-Inertial Motion

We can recover the expression (6.10) for the acceleration in a rotating (non-inertial) frame from a Lagrangian formulation as follows. The Lagrangian for a particle of mass m moving in a non-inertial rotating frame (with its origin coinciding with the fixed-frame origin) in the presence of the potential $U(\mathbf{r})$ is expressed as

$$L(\mathbf{r}, \dot{\mathbf{r}}) = \frac{m}{2}|\dot{\mathbf{r}} + \boldsymbol{\omega} \times \mathbf{r}|^2 - U(\mathbf{r}), \qquad (6.11)$$

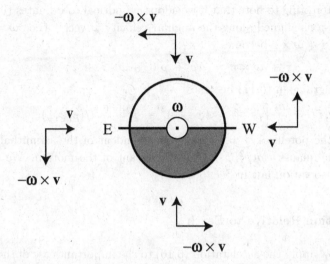

Fig. 6.2 Coriolis acceleration for a falling object on Earth (the shaded area shows the night-side of Earth and the rotation angular velocity is pointing out of the page).

where ω is the angular velocity vector and we use the formula

$$|\dot{\mathbf{r}} + \omega \times \mathbf{r}|^2 = |\dot{\mathbf{r}}|^2 + 2\,\omega \cdot (\mathbf{r} \times \dot{\mathbf{r}}) + \left[\omega^2\,r^2 - (\omega \cdot \mathbf{r})^2\right].$$

Using the Lagrangian (6.11), we now derive the general Euler-Lagrange equations for $\mathbf{r}$. First, we derive an expression for the canonical momentum

$$\mathbf{p} = \frac{\partial L}{\partial \dot{\mathbf{r}}} = m(\dot{\mathbf{r}} + \omega \times \mathbf{r}), \tag{6.12}$$

so that the time derivative of the canonical momentum is

$$\frac{d}{dt}\left(\frac{\partial L}{\partial \dot{\mathbf{r}}}\right) = m(\ddot{\mathbf{r}} + \dot{\omega} \times \mathbf{r} + \omega \times \dot{\mathbf{r}}).$$

Next, we derive the generalized force

$$\frac{\partial L}{\partial \mathbf{r}} = -\nabla U(\mathbf{r}) - m\left[\omega \times \dot{\mathbf{r}} + \omega \times (\omega \times \mathbf{r})\right],$$

so that the Euler-Lagrange equations are

$$m\,\ddot{\mathbf{r}} = -\nabla U(\mathbf{r}) - m\left[\dot{\omega} \times \mathbf{r} + 2\,\omega \times \dot{\mathbf{r}} + \omega \times (\omega \times \mathbf{r})\right]. \tag{6.13}$$

Here, the potential energy term generates the fixed-frame acceleration, $-\nabla U = m\mathbf{a}_f$, and thus the Euler-Lagrange equation (6.13) yields Eq. (6.10) for $\mathbf{a}_f = \ddot{\mathbf{r}}$.

It is interesting to note that if we adopt cylindrical coordinates (ρ, φ, z), with the z-axis aligned along the angular velocity $\boldsymbol{\omega}$ vector (i.e., $\boldsymbol{\omega} = \omega\,\widehat{\mathbf{z}}$), the vector $\dot{\mathbf{r}} + \boldsymbol{\omega} \times \mathbf{r}$ becomes

$$\dot{\mathbf{r}} + \boldsymbol{\omega} \times \mathbf{r} = \dot{\rho}\,\widehat{\rho} + \rho\,(\dot{\varphi} + \omega)\,\widehat{\varphi} + \dot{z}\,\widehat{\mathbf{z}},$$

and the Lagrangian (6.11) becomes

$$L(\mathbf{r}, \dot{\mathbf{r}}) = \frac{m}{2}\left[\dot{\rho}^2 + \rho^2\,(\dot{\varphi} + \omega)^2 + \dot{z}^2\right] - U(\rho, \varphi, z). \qquad (6.14)$$

Hence, if the potential $U = U(\rho, z)$ is independent of the azimuthal angle φ, then the quantity $m\,\rho^2(\dot{\varphi} + \omega)$ is a constant of the motion. We return to this conservation law in Sec. 6.4.2.

6.4   Motion Relative to Earth

We can now apply the acceleration (6.10) to the important case of the fixed frame of reference having its origin at the center of Earth (point O' in Fig. 6.3) and the rotating frame of reference having its origin at latitude λ and longitude ψ (point O in Fig. 6.3). We note that the rotation of the Earth is now represented as $\dot{\psi} = \omega$ (with $\dot{\boldsymbol{\omega}} = 0$).

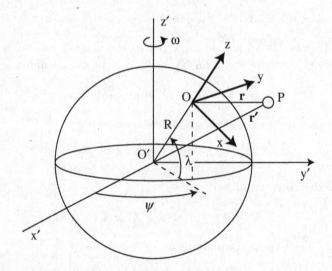

Fig. 6.3   Earth frame.

We arrange the (x, y, z)-axes of the rotating frame so that the z-axis is a continuation of the position vector $\mathbf{R}$ of the rotating-frame origin, i.e., $\mathbf{R} = R\hat{\mathbf{z}}$ in the rotating frame (where $R = 6378$ km is the radius of a *spherical* Earth). When expressed in terms of the fixed-frame latitude angle λ and the azimuthal angle ψ, the unit vector $\hat{\mathbf{z}}$ is

$$\hat{\mathbf{z}} = \cos\lambda \, (\cos\psi\,\hat{\mathbf{x}}' + \sin\psi\,\hat{\mathbf{y}}') + \sin\lambda\,\hat{\mathbf{z}}'.$$

Likewise, we choose the x-axis to be tangent to a *great circle* passing through the North and South poles, so that

$$\hat{\mathbf{x}} = \sin\lambda \, (\cos\psi\,\hat{\mathbf{x}}' + \sin\psi\,\hat{\mathbf{y}}') - \cos\lambda\,\hat{\mathbf{z}}',$$

i.e., $\hat{\mathbf{x}}$ points southward. Lastly, the y-axis is chosen such that

$$\hat{\mathbf{y}} = \hat{\mathbf{z}} \times \hat{\mathbf{x}} = -\sin\psi\,\hat{\mathbf{x}}' + \cos\psi\,\hat{\mathbf{y}}',$$

i.e., $\hat{\mathbf{y}}$ points eastward.

We now consider the acceleration of a point P as observed in the rotating frame O by writing Eq. (6.10) as

$$\frac{d^2\mathbf{r}}{dt^2} = \mathbf{g}_0 - \ddot{\mathbf{R}} - \boldsymbol{\omega} \times (\boldsymbol{\omega} \times \mathbf{r}) - 2\boldsymbol{\omega} \times \frac{d\mathbf{r}}{dt}. \tag{6.15}$$

The first term represents the *pure* gravitational acceleration due to the gravitational pull of the Earth on point P (as observed in the fixed frame located at Earth's center)

$$\mathbf{g}_0 = -\frac{GM}{|\mathbf{r}'|^3}\,\mathbf{r}', \tag{6.16}$$

where $\mathbf{r}' = \mathbf{R} + \mathbf{r}$ is the position of point P in the fixed frame and $\mathbf{r}$ is the location of P in the rotating frame. When expressed in terms of rotating-frame spherical coordinates (r, θ, φ):

$$\mathbf{r} = r\,[\,\sin\theta\,(\cos\varphi\,\hat{\mathbf{x}} + \sin\varphi\,\hat{\mathbf{y}}) + \cos\theta\,\hat{\mathbf{z}}\,],$$

the fixed-frame position $\mathbf{r}'$ is written as

$$\mathbf{r}' = (R + r\cos\theta)\,\hat{\mathbf{z}} + r\sin\theta\,(\cos\varphi\,\hat{\mathbf{x}} + \sin\varphi\,\hat{\mathbf{y}}),$$

and thus

$$|\mathbf{r}'|^3 = \left(R^2 + 2Rr\cos\theta + r^2\right)^{3/2}.$$

The pure gravitational acceleration (6.16) is, therefore, expressed in the rotating frame of the Earth as

$$\mathbf{g}_0 = -g_0 \left[\frac{(1 + \epsilon\cos\theta)\,\hat{\mathbf{z}} + \epsilon\sin\theta\,(\cos\varphi\,\hat{\mathbf{x}} + \sin\varphi\,\hat{\mathbf{y}})}{(1 + 2\epsilon\cos\theta + \epsilon^2)^{3/2}} \right] \tag{6.17}$$

$$= -g_0 \,[(1 - 2\epsilon\cos\theta)\,\hat{\mathbf{z}} + \epsilon\sin\theta\,(\cos\varphi\,\hat{\mathbf{x}} + \sin\varphi\,\hat{\mathbf{y}}) + \cdots],$$

where $g_0 = GM/R^2 = 9.789$ m/s^2 and $\epsilon = r/R \ll 1$ (e.g., $\epsilon \sim 10^{-6}$ at $r \sim 10$ m).

The angular velocity in the fixed frame is $\boldsymbol{\omega} = \omega\,\widehat{z}'$, where

$$\omega = \frac{2\pi\ \text{rad}}{24 \times 3600\ \text{sec}} = 7.27 \times 10^{-5}\ \text{rad/s}$$

is the angular rotation speed of Earth about its axis. In the rotating frame, we find

$$\boldsymbol{\omega} = \omega\,(\sin\lambda\,\widehat{z} - \cos\lambda\,\widehat{x}). \tag{6.18}$$

Because the position vector $\mathbf{R}$ rotates with the origin of the rotating frame, its time derivatives yield

$$\dot{\mathbf{R}} = \boldsymbol{\omega} \times \mathbf{R} = (\omega R \cos\lambda)\,\widehat{y},$$
$$\ddot{\mathbf{R}} = \boldsymbol{\omega} \times \dot{\mathbf{R}}_f = \boldsymbol{\omega} \times (\boldsymbol{\omega} \times \mathbf{R}) = -\omega^2 R \cos\lambda\,(\cos\lambda\,\widehat{z} + \sin\lambda\,\widehat{x}),$$

and thus the centrifugal acceleration due to $\mathbf{R}$ is

$$-\ddot{\mathbf{R}} = -\boldsymbol{\omega} \times (\boldsymbol{\omega} \times \mathbf{R}) = \alpha\,g_0\,\cos\lambda\,(\cos\lambda\,\widehat{z} + \sin\lambda\,\widehat{x}), \tag{6.19}$$

where $\omega^2 R = 0.0337$ m/s^2 can be expressed in terms of the *pure* gravitational acceleration g_0 as $\omega^2 R = \alpha\,g_0$, where $\alpha = 3.4 \times 10^{-3}$ is the normalized centrifugal acceleration. We now define the physical gravitational acceleration as

$$\mathbf{g} = \mathbf{g}_0 - \boldsymbol{\omega} \times (\boldsymbol{\omega} \times \mathbf{R})$$
$$= g_0\left[-(1 - \alpha\cos^2\lambda)\,\widehat{z} + (\alpha\cos\lambda\sin\lambda)\,\widehat{x}\right], \tag{6.20}$$

where terms of order $\epsilon = r/R$ have been neglected (since $\epsilon \ll \alpha$). For example, a plumb line experiences a small angular deviation $\delta(\lambda)$ (southward) from the true vertical given as

$$\tan\delta(\lambda) = \frac{g_x}{|g_z|} = \frac{\alpha\sin 2\lambda}{(2 - \alpha) + \alpha\cos 2\lambda}.$$

This function exhibits a maximum at a latitude $\overline{\lambda}$ defined as $\cos 2\overline{\lambda} = -\alpha/(2 - \alpha)$, so that

$$\tan\overline{\delta} = \frac{\alpha\sin 2\overline{\lambda}}{(2 - \alpha) + \alpha\cos 2\overline{\lambda}} = \frac{\alpha}{2\sqrt{1 - \alpha}} \simeq 1.7 \times 10^{-3},$$

or

$$\overline{\delta} \simeq 5.86\,\text{arcmin} \quad \text{at} \quad \overline{\lambda} \simeq \left(\frac{\pi}{4} + \frac{\alpha}{4}\right)\,\text{rad} = 45.05^\circ.$$

We now return to Eq. (6.15), which is written to lowest order in ϵ and α as

$$\frac{d^2\mathbf{r}}{dt^2} = -g\,\hat{\mathbf{z}} - 2\boldsymbol{\omega} \times \frac{d\mathbf{r}}{dt}, \tag{6.21}$$

which includes the effective (constant) gravitational acceleration g and the Coriolis acceleration

$$-2\boldsymbol{\omega} \times \frac{d\mathbf{r}}{dt} = -2\omega\,[\,(\dot{x}\sin\lambda + \dot{z}\cos\lambda)\hat{\mathbf{y}} - \dot{y}\,(\sin\lambda\,\hat{\mathbf{x}} + \cos\lambda\,\hat{\mathbf{z}})\,].$$

Thus, we find the three components of Eq. (6.21) written explicitly as

$$\left.\begin{array}{l} \ddot{x} = 2\,\omega\sin\lambda\,\dot{y} \\ \ddot{y} = -2\,\omega\,(\sin\lambda\,\dot{x} + \cos\lambda\,\dot{z}) \\ \ddot{z} = -g + 2\,\omega\cos\lambda\,\dot{y} \end{array}\right\}. \tag{6.22}$$

An interesting comment can be made concerning horizontal motion $(\dot{z} = 0)$ on Earth in the presence of the Coriolis acceleration $(\ddot{x}, \ddot{y}) = 2\omega\sin\lambda\,(\dot{y}, -\dot{x})$. By calculating the Frenet-Serret curvature [see Eq. (A.4)] for this planar motion, we find

$$\kappa \equiv \frac{\dot{y}\,\ddot{x} - \dot{x}\,\ddot{y}}{(\dot{x}^2 + \dot{y}^2)^{3/2}} = \frac{2\omega}{v}\,\sin\lambda,$$

where v is a constant. Hence, the Coriolis acceleration generates an *inertia* circle with a radius equal to $v/(2\,\omega\sin\lambda)$. For example, a particle drifting horizontally (e.g., at sea) with speed 10 cm/s at latitude $\lambda = 45°$ performs an inertia circle with a radius of approximately 1 km.

A first integration of Eq. (6.22) yields

$$\left.\begin{array}{l} \dot{x} = 2\omega\sin\lambda\,y + C_x \\ \dot{y} = -2\omega\,(\sin\lambda\,x + \cos\lambda\,z) + C_y \\ \dot{z} = -gt + 2\omega\cos\lambda\,y + C_z \end{array}\right\}, \tag{6.23}$$

where (C_x, C_y, C_z) are constants defined from initial conditions (x_0, y_0, z_0) and $(\dot{x}_0, \dot{y}_0, \dot{z}_0)$:

$$\left.\begin{array}{l} C_x = \dot{x}_0 - 2\omega\sin\lambda\,y_0 \\ C_y = \dot{y}_0 + 2\omega\,(\sin\lambda\,x_0 + \cos\lambda\,z_0) \\ C_z = \dot{z}_0 - 2\omega\cos\lambda\,y_0 \end{array}\right\}. \tag{6.24}$$

A second integration of Eq. (6.23) yields

$$x(t) = x_0 + C_x t + 2\omega\sin\lambda \int_0^t y\,dt,$$

$$y(t) = y_0 + C_y t - 2\omega\sin\lambda \int_0^t x\,dt - 2\omega\cos\lambda \int_0^t z\,dt,$$

$$z(t) = z_0 + C_z t - \frac{1}{2}g t^2 + 2\omega\cos\lambda \int_0^t y\,dt,$$

which can also be rewritten as

$$\left. \begin{array}{l} x(t) = x_0 + C_x t + \delta x(t) \\ y(t) = y_0 + C_y t + \delta y(t) \\ z(t) = z_0 + C_z t - \frac{1}{2} g t^2 + \delta z(t) \end{array} \right\}, \tag{6.25}$$

where the Coriolis *drifts* $(\delta x, \delta y, \delta z)$ are

$$\delta x(t) = 2 \omega \sin \lambda \left(y_0 t + \frac{1}{2} C_y t^2 + \int_0^t \delta y \, dt \right) \tag{6.26}$$

$$\delta y(t) = - 2 \omega \sin \lambda \left(x_0 t + \frac{1}{2} C_x t^2 + \int_0^t \delta x \, dt \right) \tag{6.27}$$

$$- 2 \omega \cos \lambda \left(z_0 t + \frac{1}{2} C_z t^2 - \frac{1}{6} g t^3 + \int_0^t \delta z \, dt \right)$$

$$\delta z(t) = 2 \omega \cos \lambda \left(y_0 t + \frac{1}{2} C_y t^2 + \int_0^t \delta y \, dt \right). \tag{6.28}$$

Note that each Coriolis drift can be expressed as an infinite series in powers of ω and that all Coriolis effects vanish when $\omega = 0$ (i.e., a fixed Earth).

6.4.1   Free-Fall Problem Revisited

As an example of the importance of the Coriolis effects in describing motion relative to Earth, we consider the simple *free-fall* problem, where

$$(x_0, y_0, z_0) = (0, 0, h) \quad \text{and} \quad (\dot{x}_0, \dot{y}_0, \dot{z}_0) = (0, 0, 0),$$

so that the constants (6.24) are $C_x = 0 = C_z$ and $C_y = 2 \omega h \cos \lambda$. Substituting these constants into Eqs. (6.25) and keeping only terms up to first order in ω in the Coriolis drifts (6.26)-(6.28), the equations (6.25) become

$$x(t) = 0, \tag{6.29}$$

$$y(t) = \frac{1}{3} g t^3 \omega \cos \lambda, \tag{6.30}$$

$$z(t) = h - \frac{1}{2} g t^2. \tag{6.31}$$

Hence, a free-falling object starting from rest at height h touches the ground $z(T) = 0$ after a time $T = \sqrt{2h/g}$ after which time the object has drifted eastward by a distance of

$$y(T) = \frac{1}{3} g T^3 \omega \cos \lambda = \frac{\omega \cos \lambda}{3} \sqrt{\frac{8h^3}{g}}.$$

This eastward Coriolis drift is maximum at the equator ($\lambda = 0$). At a height of 100 m and latitude 45°, for example, we find an eastward drift of 15.5 mm.

6.4.2 *Foucault Pendulum*

In 1851, Jean Bernard Léon Foucault (1819-1868) was able to demonstrate, in a classic experiment demonstrating Earth's rotation, the role played by Coriolis effect in his investigations of the motion of a pendulum (of length ℓ and mass m) in the rotating frame of the Earth. His analysis showed that, because of the Coriolis acceleration associated with the rotation of the Earth, the motion of the pendulum exhibits a precession motion whose period depends on the latitude at which the pendulum is located.

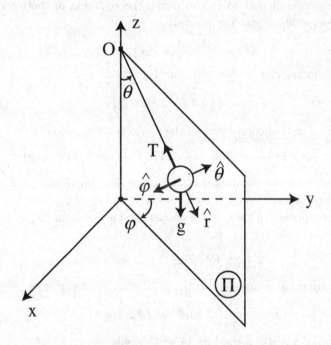

Fig. 6.4   Plane of oscillation of the Foucault pendulum.

The equation of motion for the pendulum is given as

$$\ddot{\mathbf{r}} = \mathbf{a}_f - 2\,\boldsymbol{\omega} \times \dot{\mathbf{r}}, \qquad (6.32)$$

where $\mathbf{a}_f = \mathbf{g} + \mathbf{T}/m$ is the net fixed-frame acceleration of the pendulum expressed in terms of the gravitational acceleration **g** and the string tension **T** (see Fig. 6.4). Note that the vectors **g** and **T** span a plane Π in which the pendulum moves in the absence of the Coriolis acceleration $- 2\boldsymbol{\omega} \times \dot{\mathbf{r}}$. Using spherical coordinates (r, θ, φ) in the rotating frame and placing the origin O of the pendulum system at its pivot point (see Fig. 6.4), the position of the pendulum bob is

$$\mathbf{r} = \ell\,[\,\sin\theta\,(\sin\varphi\,\widehat{\mathbf{x}} + \cos\varphi\,\widehat{\mathbf{y}}) - \cos\theta\,\widehat{\mathbf{z}}\,] = \ell\,\widehat{r}(\theta, \varphi). \tag{6.33}$$

From this definition, we construct the unit vectors $\widehat{\theta}$ and $\widehat{\varphi}$ as

$$\frac{\partial \widehat{r}}{\partial \theta} = \widehat{\theta}, \quad \frac{\partial \widehat{r}}{\partial \varphi} = \sin\theta\,\widehat{\varphi}, \quad \text{and} \quad \frac{\partial \widehat{\theta}}{\partial \varphi} = \cos\theta\,\widehat{\varphi}. \tag{6.34}$$

Note that, whereas the unit vectors $\widehat{r}$ and $\widehat{\theta}$ lie on the plane Π, the unit vector $\widehat{\varphi}$ is perpendicular to it and, thus, the equation of motion of the pendulum *perpendicular* to the plane Π is

$$\ddot{\mathbf{r}} \cdot \widehat{\varphi} = - 2\,(\boldsymbol{\omega} \times \dot{\mathbf{r}}) \cdot \widehat{\varphi}. \tag{6.35}$$

The pendulum velocity is obtained from Eq. (6.33) as

$$\dot{\mathbf{r}} = \ell\left(\dot{\theta}\,\widehat{\theta} + \dot{\varphi}\,\sin\theta\,\widehat{\varphi}\right), \tag{6.36}$$

so that the azimuthal component of the Coriolis acceleration is

$$- 2\,(\boldsymbol{\omega} \times \dot{\mathbf{r}}) \cdot \widehat{\varphi} = 2\,\ell\omega\,\dot{\theta}\,(\sin\lambda\,\cos\theta + \cos\lambda\,\sin\theta\,\sin\varphi).$$

If the length ℓ of the pendulum is large, the angular deviation θ of the pendulum can be small enough that $\sin\theta \ll 1$ and $\cos\theta \simeq 1$ and, thus, the azimuthal component of the Coriolis acceleration is approximately (ignoring $\dot{\theta}\,\sin\theta$)

$$- 2\,(\boldsymbol{\omega} \times \dot{\mathbf{r}}) \cdot \widehat{\varphi} \simeq 2\,\ell\,(\omega\,\sin\lambda)\,\dot{\theta}. \tag{6.37}$$

Next, the azimuthal component of the pendulum acceleration is

$$\ddot{\mathbf{r}} \cdot \widehat{\varphi} = \ell\left(\ddot{\varphi}\,\sin\theta + 2\,\dot{\theta}\dot{\varphi}\,\cos\theta\right),$$

which for small angular deviations ($\theta \ll 1$) yields

$$\ddot{\mathbf{r}} \cdot \widehat{\varphi} \simeq 2\,\ell\,(\dot{\varphi})\,\dot{\theta}. \tag{6.38}$$

By combining these expressions into Eq. (6.35), we obtain an expression for the precession angular frequency of the Foucault pendulum

$$\dot{\varphi} = \omega\,\sin\lambda \tag{6.39}$$

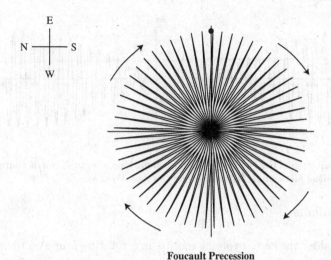

Foucault Precession

Fig. 6.5   Numerical solution of the Foucault pendulum. The precession motion is clockwise and the initial plane of oscillation is vertical (East-West axis).

as a function of latitude λ. As expected, the precession motion is clockwise in the Northern Hemisphere and reaches a maximum at the North Pole $(\lambda = 90^o)$. Note that the precession period of the Foucault pendulum is $(1\ \text{day}/\sin\lambda)$ so that the period is 1.41 days at a latitude of 45° or 2 days at a latitude of 30°.

The more traditional approach to describing the precession motion of the Foucault pendulum makes use of Cartesian coordinates (x, y, z). The motion of the Foucault pendulum in the (x, y)-plane is described in terms of Eqs. (6.32) as

$$\left.\begin{array}{c} \ddot{x} + \omega_0^2\, x = 2\,\omega \sin\lambda\,\dot{y} \\ \ddot{y} + \omega_0^2\, y = -\,2\,\omega \sin\lambda\,\dot{x} \end{array}\right\}, \qquad (6.40)$$

where $\omega_0^2 = T/m\ell \simeq g/\ell$ and $\dot{z} \simeq 0$ if ℓ is very large. Figure 6.5 shows the numerical solution of Eqs. (6.40) for the Foucault pendulum starting from rest at $(x_0, y_0) = (0, 1)$ with $2\,(\omega/\omega_0)\sin\lambda = 0.05$ at $\lambda = 45^o$. Figure 6.6 shows that, over a finite period of time, the pendulum motion progressively moves from the East-West axis to the North-South axis.

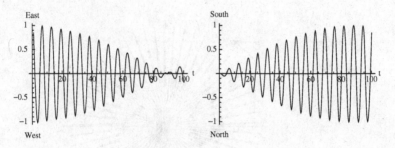

Fig. 6.6   Projection of the Foucault pendulum along East-West and North-South directions. The initial plane of oscillation is along the East-West axis.

6.5   Problems

1. (a) Consider the case involving motion in a rotating frame on the (x, y)-plane perpendicular to the angular velocity vector $\boldsymbol{\omega} = \omega\,\widehat{\mathbf{z}}$ with the potential energy

$$U(\mathbf{r}) \;=\; \frac{1}{2}\, k\, \left(x^2 + y^2\right).$$

Using the Euler-Lagrange equations (6.13), derive the equations of motion for x and y.

(b) By using the equations of motion derived in Part (a), show that the canonical angular momentum $\ell = \widehat{\mathbf{z}} \cdot (\mathbf{r} \times \mathbf{p})$ is a constant of the motion.

2. If a particle is projected vertically upward to a height h above a point on the Earth's surface at a northern latitude λ, show that it strikes the ground at a point

$$\frac{4\omega}{3}\, \cos\lambda \, \sqrt{\frac{8\,h^3}{g}}$$

to the west.   (Neglect air resistance, and consider only small vertical heights.)

3. For the potential

$$U(\mathbf{r}, \dot{\mathbf{r}}) \;=\; V(r) + \boldsymbol{\sigma} \cdot \mathbf{r} \times \dot{\mathbf{r}},$$

where $V(r)$ denotes an arbitrary central potential and $\boldsymbol{\sigma}$ denotes an arbitrary constant vector, derive the Euler-Lagrange equations of motion in terms of spherical coordinates.

4. The Lagrangian for the Foucault-pendulum equations (6.40) is
$$L(x, y; \dot{x}, \dot{y}) = \frac{1}{2}\left(\dot{x}^2 + \dot{y}^2\right) - \frac{\omega_0^2}{2}\left(x^2 + y^2\right) + \omega \sin\lambda\,(x\,\dot{y} - \dot{x}\,y).$$

(a) By using the polar transformation
$$x(t) = \rho(t)\,\cos\varphi(t) \quad \text{and} \quad y(t) = \rho(t)\,\sin\varphi(t),$$
derive the new Lagrangian $L(\rho; \dot{\rho}, \dot{\varphi})$.

(b) Since the new Lagrangian $L(\rho; \dot{\rho}, \dot{\varphi})$ is independent of φ, derive an expression for the conserved momentum p_φ and find the Routhian $R(\rho, \dot{\rho}; p_\varphi)$ and the Routh-Euler-Lagrangian equation for ρ.

5. We define the complex-valued function $q = y + i\,x = \ell\,\sin\theta\,e^{i\varphi}$, so that Eq. (6.40) becomes
$$\ddot{q} + \omega_0^2\,q - 2i\,\omega \sin\lambda\,\dot{q} = 0.$$

(a) Insert the eigenfunction $q(t) = \rho\exp(i\Omega t)$ into this equation and find that the solution for the eigenfrequency Ω is
$$\Omega = \omega\sin\lambda \pm \sqrt{\omega^2 \sin^2\lambda + \omega_0^2},$$
so that the eigenfunction is
$$q = \rho\,e^{i\omega\,\sin\lambda t}\,\sin\left(\sqrt{\omega^2 \sin^2\lambda + \omega_0^2}\,t\right).$$

(b) Verify that
$$\rho\sin\left(\sqrt{\omega^2 \sin^2\lambda + \omega_0^2}\,t\right) = \ell\,\sin\theta \simeq \ell\,\theta(t),$$
and
$$\varphi(t) = (\omega\sin\lambda)\,t,$$
from which we recover the Foucault pendulum precession frequency (6.39).

6. A projectile is fired at latitude λ with velocity v_0 directed toward the west at angle α with respect to the horizontal. In what follows, we neglect terms of order ω^2 in the Coriolis drift equations (6.25).

(a) Show that the time taken to reach maximum height is
$$\frac{v_0\sin\alpha}{g}\left(1 - 2\,\frac{\omega v_0}{g}\cos\lambda\cos\alpha\right).$$

(b) Show that the maximum height reached is

$$\frac{v_0^2 \sin \alpha^2}{2g} \left(1 \, - \, 4 \, \frac{\omega \, v_0}{g} \, \cos \lambda \, \cos \alpha \right).$$

(c) Show that the maximum range of the projectile is

$$\frac{v_0^2}{g} \left[\sin 2\alpha \, + \, \frac{\omega \, v_0}{3 \, g} \, \cos \lambda \, \sin \alpha \left(8 \sin^2 \alpha - 6 \right) \right].$$

Hence, the maximum range is larger than the case where $\omega = 0$ if $\alpha > 60^\circ$.

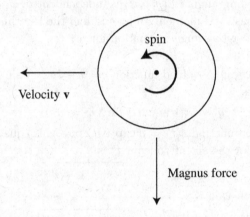

Fig. 6.7 Magnus force acting on a spinning sphere (with spin axis directed out of the page) moving with velocity **v**.

7. The equations of motion for a sphere of mass m traveling in air with velocity **v** in a constant gravitational field **g** are

$$m \, \frac{d\mathbf{v}}{dt} \, = \, m\mathbf{g} \, + \, \mu \, \boldsymbol{\omega} \times \mathbf{v} \, - \, (\beta \, \mathbf{v} + \gamma \, v \mathbf{v}), \tag{6.41}$$

where the second term is the Magnus force (see Fig. 6.7), and the last two terms represent the effects of linear and quadratic air resistance. The air-resistance coefficients for a sphere of diameter D are $\beta = 1.6 \times 10^{-4} \times D$ and $\gamma = 0.25 \times D^2$ in SI units. The Magnus coefficient for a sphere of diameter D traveling in air (with density 1.168 kg/m^3 at 25 °C and 1 atm) is $\mu = (\pi^2/8) \, 1.168 \times D^3$. Previous studies have shown that the torque experienced by the sphere during its trajectory is negligible and, thus, the angular velocity $\boldsymbol{\omega} = \omega \, \hat{\boldsymbol{\omega}}$ is treated as constant in Eq. (6.41).

(a) Show that the Magnus force is energy-conserving (i.e., it does no work on the sphere) and, thus, its sole purpose is to change the direction of motion of the sphere.

(b) For a sphere of diameter $D = 0.07$ m and mass $m = 0.145$ kg (e.g., a baseball) spinning at $\omega = 30$ rev/sec and traveling horizontally at $v = 44$ m/s, compare the magnitudes of the Magnus force (assume that $\boldsymbol{\omega} \perp \mathbf{v}$) and the linear and quadratic air-resistance drag forces with the sphere's weight.

8. To analyze the three-dimensional motion of a baseball described by Eq. (6.41), we use Cartesian coordinates (x, y, z) with the origin located at the pitcher's mound, the x-axis is directed toward home-plate (located approximately 18 m away), the y-axis is directed toward first base, and the z-axis is directed upward (i.e., $\mathbf{g} = -g\hat{\mathbf{z}}$). The standard pitches in the arsenal of a baseball pitcher are the fast-ball ($\hat{\boldsymbol{\omega}} = -\hat{\mathbf{y}}$), the curve-ball ($\hat{\boldsymbol{\omega}} = \hat{\mathbf{y}}$), and the slider ($\hat{\boldsymbol{\omega}} = \hat{\mathbf{z}}$).

(a) In general, we may write the rotation unit vector $\hat{\boldsymbol{\omega}}$ for fast-balls, curve-balls, and sliders in terms of an angle ϕ in the (y, z)-plane as $\hat{\boldsymbol{\omega}} = \cos\phi\,\hat{\mathbf{y}} + \sin\phi\,\hat{\mathbf{z}}$, where $\phi = \pi$ for a fast-ball, $\phi = \pi/2$ for a slider, and $\phi = 0$ for a curve-ball. Find the general expression for the Magnus-force vector component $\hat{\boldsymbol{\omega}} \times \mathbf{v}$ for these standard pitches.

(b) For a fast-ball and a curve-ball, i.e., the front end of the baseball is rotating upward (downward) for a fast-ball (curve-ball), determine the direction of the Magnus-force vector component $\hat{\boldsymbol{\omega}} \times \mathbf{v}$ and discuss qualitatively which ball appears to rise or sink. In addition, if the fast-ball (curve-ball) is initially released horizontally ($\mathbf{v}_0 = v_0\,\hat{\mathbf{x}}$), discuss how its downward motion under the action of gravity ($\dot{z} < 0$) causes it to accelerate (decelerate) along the x-axis under the action of the Magnus force.

(c) Discuss qualitatively the dominant Magnus-force effect on a slider.

Chapter 7

Rigid Body Motion

So far objects have been considered as point-like particles. In the present Chapter, we consider objects known as *rigid bodies* defined as non-deformable discrete collections of massive particles or continuous mass distributions. The inertial properties of such objects are described not only in terms of their masses (i.e., their translational inertia) but also in terms of how their masses are distributed about their instantaneous axis of rotation (i.e., their rotational inertia).

7.1 Inertia Tensor of a Rigid Body

7.1.1 *Discrete Particle Distribution*

We begin our description of rigid body motion by considering the case of a rigid *discrete* particle distribution in which the inter-particle distances are constant. The position of each particle α as measured from a *fixed* laboratory (LAB) frame is

$$\mathbf{r}'_\alpha = \mathbf{R} + \mathbf{r}_\alpha,$$

where

$$\mathbf{R} \equiv \frac{\sum_\alpha m_\alpha \mathbf{r}'_\alpha}{\sum_\alpha m_\alpha} \equiv \sum_\alpha \frac{m_\alpha}{M} \mathbf{r}'_\alpha$$

is the position of the center of mass (CM) in the LAB ($M \equiv \sum_\alpha m_\alpha$ is the total mass of the system) and $\mathbf{r}_\alpha$ is the position of the particle in the CM frame (i.e., $\sum_\alpha m_\alpha \mathbf{r}_\alpha \equiv 0$).

The velocity of particle α in the LAB frame is

$$\mathbf{v}'_\alpha = \dot{\mathbf{R}} + \boldsymbol{\omega} \times \mathbf{r}_\alpha, \tag{7.1}$$

159

where $\boldsymbol{\omega}$ is the angular velocity vector associated with the rotation of the particle distribution about an axis of rotation which passes through the CM, the velocity $\mathbf{v}_\alpha = 0$ for each particle of a discrete rigid body, and $\dot{\mathbf{R}}$ is the CM velocity in the LAB frame. The total linear momentum in the LAB frame is equal to the momentum of the center of mass since

$$\mathbf{P}' = \sum_\alpha m_\alpha \mathbf{v}'_\alpha = M\dot{\mathbf{R}} + \boldsymbol{\omega} \times \left(\sum_\alpha m_\alpha \mathbf{r}_\alpha \right) = M\dot{\mathbf{R}},$$

i.e., the total momentum of a rigid body in its CM frame is zero.

Next, the total angular momentum in the LAB frame is expressed as

$$\mathbf{L}' = \sum_\alpha m_\alpha \mathbf{r}'_\alpha \times \mathbf{v}'_\alpha = M\,\mathbf{R} \times \dot{\mathbf{R}} + \sum_\alpha m_\alpha \mathbf{r}_\alpha \times (\boldsymbol{\omega} \times \mathbf{r}_\alpha), \quad (7.2)$$

and the kinetic energy of particle α (with mass m_α) in the LAB frame is

$$K'_\alpha = \frac{m_\alpha}{2} |\mathbf{v}'_\alpha|^2 = \frac{m_\alpha}{2} \left(|\dot{\mathbf{R}}|^2 + 2\,\dot{\mathbf{R}} \cdot \boldsymbol{\omega} \times \mathbf{r}_\alpha + |\boldsymbol{\omega} \times \mathbf{r}_\alpha|^2 \right).$$

The total kinetic energy $K' = \sum_\alpha K'_\alpha$ of the particle distribution is thus

$$K' = \frac{M}{2} |\dot{\mathbf{R}}|^2 + \frac{1}{2} \left\{ \omega^2 \left(\sum_\alpha m_\alpha r_\alpha^2 \right) - \boldsymbol{\omega}\boldsymbol{\omega} : \left(\sum_\alpha m_\alpha \mathbf{r}_\alpha \mathbf{r}_\alpha \right) \right\}. \tag{7.3}$$

Looking at Eqs. (7.2) and (7.3), we introduce the *inertia tensor* of the particle distribution

$$\mathbf{I} = \sum_\alpha m_\alpha \left(r_\alpha^2 \,\underline{1} - \mathbf{r}_\alpha \mathbf{r}_\alpha \right), \tag{7.4}$$

where $\underline{1}$ denotes the unit tensor (i.e., in Cartesian coordinates, $\underline{1} = \widehat{\mathbf{x}}\widehat{\mathbf{x}} + \widehat{\mathbf{y}}\widehat{\mathbf{y}} + \widehat{\mathbf{z}}\widehat{\mathbf{z}}$). In terms of the inertia tensor (7.4), the angular momentum of a rigid body in the CM frame and its rotational kinetic energy are

$$\mathbf{L} = \mathbf{I} \cdot \boldsymbol{\omega} \quad \text{and} \quad K_{rot} = \frac{1}{2}\,\boldsymbol{\omega} \cdot \mathbf{I} \cdot \boldsymbol{\omega}. \tag{7.5}$$

The inertia tensor (7.4) can also be represented as a matrix

$$\mathbf{I} = \begin{pmatrix} \sum_\alpha m_\alpha \left(y_\alpha^2 + z_\alpha^2 \right) & -\sum_\alpha m_\alpha \left(x_\alpha y_\alpha \right) & -\sum_\alpha m_\alpha \left(x_\alpha z_\alpha \right) \\ -\sum_\alpha m_\alpha \left(y_\alpha x_\alpha \right) & \sum_\alpha m_\alpha \left(x_\alpha^2 + z_\alpha^2 \right) & -\sum_\alpha m_\alpha \left(y_\alpha z_\alpha \right) \\ -\sum_\alpha m_\alpha \left(z_\alpha x_\alpha \right) & -\sum_\alpha m_\alpha \left(z_\alpha y_\alpha \right) & \sum_\alpha m_\alpha \left(x_\alpha^2 + y_\alpha^2 \right) \end{pmatrix},$$

where the symmetry property of the inertia tensor ($I^{ji} = I^{ij}$) is readily apparent.

7.1.2 Parallel-Axes Theorem

A translation of the origin from which the inertia tensor is calculated leads to a different inertia tensor. Let $\mathbf{Q}_\alpha$ denote the position of particle α in a new frame of reference and let $\boldsymbol{\rho} = \mathbf{r}_\alpha - \mathbf{Q}_\alpha$ is the displacement from point CM to point P. The new inertia tensor

$$\mathbf{J} = \sum_\alpha m_\alpha \left(Q_\alpha^2\, \underline{1} - \mathbf{Q}_\alpha \mathbf{Q}_\alpha \right)$$

can be expressed as

$$\mathbf{J} = \sum_\alpha m_\alpha \left(\rho^2\, \underline{1} - \boldsymbol{\rho}\boldsymbol{\rho} \right) + \sum_\alpha m_\alpha \left(r_\alpha^2\, \underline{1} - \mathbf{r}_\alpha \mathbf{r}_\alpha \right)$$

$$+ \left\{ \boldsymbol{\rho} \left(\sum_\alpha m_\alpha \mathbf{r}_\alpha \right) + \left(\sum_\alpha m_\alpha \mathbf{r}_\alpha \right) \boldsymbol{\rho} \right\}$$

$$- \left\{ \boldsymbol{\rho} \cdot \left(\sum_\alpha m_\alpha \mathbf{r}_\alpha \right) \right\} \underline{1}.$$

Since $M = \sum_\alpha m_\alpha$ and $\sum_\alpha m_\alpha \mathbf{r}_\alpha = 0$, we find

$$\mathbf{J} = M \left(\rho^2\, \underline{1} - \boldsymbol{\rho}\boldsymbol{\rho} \right) + \mathbf{I}_{CM}. \tag{7.6}$$

Hence, once the inertia tensor $\mathbf{I}_{CM}$ is calculated in the CM frame, it can be calculated anywhere else. Eq. (7.6) is known as the Parallel-Axes Theorem.

7.1.3 Continuous Particle Distribution

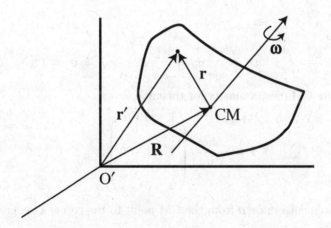

Fig. 7.1 Continuous distribution of mass.

For a continuous particle distribution the inertia tensor (7.4) becomes

$$\mathbf{I} = \int dm \left(r^2 \, \mathbf{1} - \mathbf{rr} \right),$$

(7.7)

where $dm(\mathbf{r}) = \rho(\mathbf{r}) \, d^3r$ is the infinitesimal mass element at point $\mathbf{r}$, with mass density $\rho(\mathbf{r})$.

Consider, for example, the case of a uniform cube of mass M and volume b^3, with $dm = (M/b^3) \, dx \, dy \, dz$. The inertia tensor (7.7) in the LAB frame (with the origin placed at one of its corners) has the components

$$J^{11} = \frac{M}{b^3} \int_0^b dx \int_0^b dy \int_0^b dz \cdot (y^2 + z^2) = \frac{2}{3} M b^2 = J^{22} = J^{33}$$

$$J^{12} = -\frac{M}{b^3} \int_0^b dx \int_0^b dy \int_0^b dz \cdot xy = -\frac{1}{4} M b^2 = J^{23} = J^{31}$$

and thus the inertia matrix for the uniform cube is

$$\mathbf{J} = \frac{M b^2}{12} \begin{pmatrix} 8 & -3 & -3 \\ -3 & 8 & -3 \\ -3 & -3 & 8 \end{pmatrix}.$$

(7.8)

On the other hand, the inertia tensor calculated in the CM frame (computed with axes parallel to the axes of the cube) has the components

$$I^{11} = \frac{M}{b^3} \int_{-b/2}^{b/2} dx \int_{-b/2}^{b/2} dy \int_{-b/2}^{b/2} dz \cdot (y^2 + z^2)$$

$$= \frac{1}{6} M b^2 = I^{22} = I^{33}$$

$$I^{12} = -\frac{M}{b^3} \int_{-b/2}^{b/2} dx \int_{-b/2}^{b/2} dy \int_{-b/2}^{b/2} dz \cdot xy = 0 = I^{23} = I^{31}$$

and thus the CM inertia matrix for the cube is

$$\mathbf{I} = \frac{M b^2}{6} \begin{pmatrix} 1 & 0 & 0 \\ 0 & 1 & 0 \\ 0 & 0 & 1 \end{pmatrix}.$$

(7.9)

The displacement vector ρ from the CM point to the corner O is given as

$$\rho = -\frac{b}{2} \left(\widehat{x} + \widehat{y} + \widehat{z} \right),$$

so that $\rho^2 = 3b^2/4$. By using the Parallel-Axis Theorem (7.6), the inertia tensor

$$M\left(\rho^2\,\underline{1}\,-\,\rho\rho\right)\,=\,\frac{M\,b^2}{4}\begin{pmatrix} 2 & -1 & -1 \\ -1 & 2 & -1 \\ -1 & -1 & 2 \end{pmatrix}$$

when added to the CM inertia tensor (7.9), yields the inertia tensor (7.8)

7.1.4 *Principal Axes of Inertia*

In general, the CM inertia tensor $\mathbf{I}$ can be transformed into a *diagonal* tensor with components given by the eigenvalues I_1, I_2, and I_3 of the inertia tensor. These components (known as principal moments of inertia) are the three roots of the cubic polynomial

$$I^3 \,-\, \mathrm{Tr}(\mathbf{I})\,I^2 \,+\, \mathrm{Ad}(\mathbf{I})\,I \,-\, \mathrm{Det}(\mathbf{I}) \,=\, 0, \tag{7.10}$$

obtained from $\mathrm{Det}(\mathbf{I} - I\,\underline{1}) = 0$, with coefficients

$$\mathrm{Tr}(\mathbf{I}) = I^{11} + I^{22} + I^{33},$$

$$\mathrm{Ad}(\mathbf{I}) = \mathrm{ad}_{11} + \mathrm{ad}_{22} + \mathrm{ad}_{33},$$

$$\mathrm{Det}(\mathbf{I}) = I^{11}\,\mathrm{ad}_{11} \,-\, I^{12}\,\mathrm{ad}_{12} \,+\, I^{13}\,\mathrm{ad}_{13},$$

where ad_{ij} is the determinant of the two-by-two matrix obtained from $\mathbf{I}$ by removing the ith-row and j^{th}-column from the inertia matrix $\mathbf{I}$. (See Appendix A for additional details on Linear Algebra.)

Each principal moment of inertia I_i represents the moment of inertia calculated about the principal axis of inertia with unit vector $\widehat{e}_i$. The unit vectors $(\widehat{e}_1, \widehat{e}_2, \widehat{e}_3)$ form a new frame of reference known as the *Body* frame. The unit vectors $(\widehat{e}_1, \widehat{e}_2, \widehat{e}_3)$ are related by a sequence of rotations to the Cartesian CM unit vectors $(\widehat{x}^1, \widehat{x}^2, \widehat{x}^3)$ by the relation

$$\widehat{e}_i \,=\, R_{ij}\,\widehat{x}^j, \tag{7.11}$$

where R_{ij} are components of the rotation matrix R. Note that a general rotation matrix has the form

$$\mathsf{R}_n(\alpha) \,=\, \widehat{n}\widehat{n} \,+\, \cos\alpha\,(\underline{1} \,-\, \widehat{n}\widehat{n}) \,-\, \sin\alpha\,\widehat{n} \times \underline{1}, \tag{7.12}$$

Table 7.1   Three categories of rigid bodies.

Rigid Body	Principal Moments of Inertia	Example
Asymmetric Top	$I_1 > I_2 > I_3$	textbook
Symmetric Top	$I_1 = I_2 > I_3$	oblate spheroid (pancake)
	$I_1 = I_2 < I_3$	prolate spheroid (football)
Spherical Top	$I_1 = I_2 = I_3$	cube

where the unit vector $\hat{n}$ defines the axis of rotation about which an angular rotation of angle α is performed according to the right-hand-rule. The general rotation matrix (7.12) has the following properties. First, the matrix $\mathsf{R}_n(-\alpha)$ is the inverse matrix of $\mathsf{R}_n(\alpha)$, i.e., $\mathsf{R}_n(-\alpha) \cdot \mathsf{R}_n(\alpha) = \underline{1}$. Next, the determinant of $\mathsf{R}_n(\alpha)$ is $+1$ and the eigenvalues of $\mathsf{R}_n(\alpha)$ are $+1$ and $\exp(\pm i\alpha)$ (see Appendix A.2 for further details).

A rigid body can be classified into one of three different categories (see Table 7.1) depending on its principal moments of inertia (I_1, I_2, I_3). By denoting as $\mathbf{I}'$ the diagonal inertia tensor calculated in the *body* frame of reference (along the principal axes), we find

$$\mathbf{I}' = \mathsf{R} \cdot \mathbf{I} \cdot \mathsf{R}^\top = \begin{pmatrix} I_1 & 0 & 0 \\ 0 & I_2 & 0 \\ 0 & 0 & I_3 \end{pmatrix}, \qquad (7.13)$$

where $\mathsf{R}^\top$ denotes the transpose of R, i.e., $(\mathsf{R}^\top)_{ij} = R_{ji}$. In the body frame, the inertia tensor is, therefore, expressed in dyadic form as

$$\mathbf{I}' = I_1\,\hat{e}_1\,\hat{e}_1 + I_2\,\hat{e}_2\,\hat{e}_2 + I_3\,\hat{e}_3\,\hat{e}_3, \qquad (7.14)$$

and the rotational kinetic energy (7.5) is

$$K'_{rot} = \frac{1}{2}\,\boldsymbol{\omega} \cdot \mathbf{I}' \cdot \boldsymbol{\omega} = \frac{1}{2}\left(I_1\,\omega_1^2 + I_2\,\omega_2^2 + I_3\,\omega_3^2\right). \qquad (7.15)$$

Before proceeding further, we consider the example of a dumbbell composed of two equal point masses m placed at the ends of a massless rod of total length $2\,b$ and rotating about the z-axis with angular frequency ω. Here, the positions of the two masses are expressed as

$$\mathbf{r}_\pm = \pm b\left[\sin\theta\,(\cos\varphi\,\hat{x} + \sin\varphi\,\hat{y}) + \cos\theta\,\hat{z}\,\right],$$

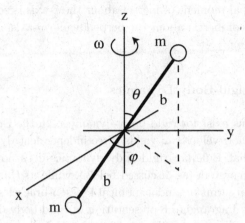

Fig. 7.2 Dumbbell with total mass $M = 2m$ and length $2b$.

so that the CM inertia tensor is $\mathbf{I} = 2\,m\,b^2\,\overline{\mathbf{I}}$, where

$$\overline{\mathbf{I}} = \begin{pmatrix} 1 - \cos^2\varphi\,\sin^2\theta & -\cos\varphi\,\sin\varphi\,\sin^2\theta & -\cos\varphi\,\cos\theta\,\sin\theta \\ -\cos\varphi\,\sin\varphi\,\sin^2\theta & 1 - \sin^2\varphi\,\sin^2\theta & -\sin\varphi\,\cos\theta\,\sin\theta \\ -\cos\varphi\,\cos\theta\,\sin\theta & -\sin\varphi\,\cos\theta\,\sin\theta & 1 - \cos^2\theta \end{pmatrix}.$$

(7.16)

After some tedious algebra, we find $\mathrm{Tr}(\mathbf{I}) = 4\,mb^2$, $\mathrm{Ad}(\mathbf{I}) = (2\,mb^2)^2$, and $\mathrm{Det}(\mathbf{I}) = 0$, and thus the cubic polynomial (7.10) has the single root $I_3 = 0$ and the double root $I_1 = I_2 = 2\,mb^2$, which makes the dumbbell a symmetric top.

The root $I_3 = 0$ clearly indicates that one of the three principal axes is the axis of symmetry of the dumbbell ($\hat{e}_3 = \hat{r}$). The other two principal axes are located on the plane perpendicular to the symmetry axis (i.e., $\hat{e}_1 = \hat{\theta}$ and $\hat{e}_2 = \hat{\varphi}$). From these principal axes, we easily recover the rotation matrix

$$\mathsf{R} = \mathsf{R}_2(-\theta) \cdot \mathsf{R}_3(\varphi) = \begin{pmatrix} \cos\varphi\,\cos\theta & \sin\varphi\,\cos\theta & -\sin\theta \\ -\sin\varphi & \cos\varphi & 0 \\ \cos\varphi\,\sin\theta & \sin\varphi\,\sin\theta & \cos\theta \end{pmatrix},$$

so that, using the spherical coordinates (r, θ, φ), we find

$$\hat{e}_1 = \cos\theta\,(\cos\varphi\,\hat{x} + \sin\varphi\,\hat{y}) - \sin\theta\,\hat{z} = \hat{\theta},$$
$$\hat{e}_2 = -\sin\varphi\,\hat{x} + \cos\varphi\,\hat{y} = \hat{\varphi},$$
$$\hat{e}_3 = \sin\theta\,(\cos\varphi\,\hat{x} + \sin\varphi\,\hat{y}) + \cos\theta\,\hat{z} = \hat{r}.$$

Indeed, the principal moment of inertia about the $\widehat{r}$-axis is zero, while the principal moments of inertia about the perpendicular $\widehat{\theta}$-axis and $\widehat{\varphi}$-axis are equally given as $2\,mb^2$.

7.2 Eulerian Rigid-Body Dynamics

Two representations exist for rigid-body dynamics. In the Eulerian representation, the angular velocity $\boldsymbol{\omega}$ yields three independent dynamical variables. We note that Eulerian rigid-body dynamics does not represent a regular Lagrangian system (as discussed below), while its Hamiltonian formulation is given in terms of a noncanonical Poisson-bracket structure (see problem 8). In the Lagrangian representation of rigid-body dynamics, the Lagrangian function is expressed in terms of three Eulerian angles and their velocities (this representation is discussed in the next Section).

7.2.1 *Euler Equations*

The time derivative of the angular momentum $\mathbf{L} = \mathbf{I} \cdot \boldsymbol{\omega}$ in the fixed (LAB) frame is given as

$$\left(\frac{d\mathbf{L}}{dt}\right)_f = \left(\frac{d\mathbf{L}}{dt}\right)_r + \boldsymbol{\omega} \times \mathbf{L} = \mathbf{N},$$

where $\mathbf{N}$ represents the external torque applied to the system (in the LAB frame) and $(d\mathbf{L}/dt)_r$ denotes the rate of change of $\mathbf{L}$ in the rotating frame. By choosing the body frame as the rotating frame, we find

$$\left(\frac{d\mathbf{L}}{dt}\right)_r = \mathbf{I} \cdot \dot{\boldsymbol{\omega}} = (I_1\,\dot{\omega}_1)\widehat{\mathbf{e}}_1 + (I_2\,\dot{\omega}_2)\widehat{\mathbf{e}}_2 + (I_3\,\dot{\omega}_3)\widehat{\mathbf{e}}_3, \qquad (7.17)$$

while

$$\boldsymbol{\omega} \times \mathbf{L} = -\,\widehat{\mathbf{e}}_1 \left[\omega_2\,\omega_3\,(I_2 - I_3) \right] - \widehat{\mathbf{e}}_2 \left[\omega_3\,\omega_1\,(I_3 - I_1) \right]$$
$$-\,\widehat{\mathbf{e}}_3 \left[\omega_1\,\omega_2\,(I_1 - I_2) \right]. \qquad (7.18)$$

Thus the time evolution of the angular momentum in the body frame of reference is described in terms of

$$\left. \begin{array}{l} I_1\,\dot{\omega}_1 - \omega_2\,\omega_3\,(I_2 - I_3) = N_1 \\ I_2\,\dot{\omega}_2 - \omega_3\,\omega_1\,(I_3 - I_1) = N_2 \\ I_3\,\dot{\omega}_3 - \omega_1\,\omega_2\,(I_1 - I_2) = N_3 \end{array} \right\}, \qquad (7.19)$$

which are known as the Euler equations for rigid-body motion. We note that the rate of change of the rotational kinetic energy (7.5) is expressed as

$$\frac{dK_{rot}}{dt} = \boldsymbol{\omega} \cdot \mathbf{I} \cdot \dot{\boldsymbol{\omega}} = \boldsymbol{\omega} \cdot (-\boldsymbol{\omega} \times \mathbf{L} + \mathbf{N}) = \mathbf{N} \cdot \boldsymbol{\omega}. \qquad (7.20)$$

In the absence of external torque ($\mathbf{N} = 0$), not only is the kinetic energy conserved but also the squared angular momentum $L^2 = \sum_{i=1}^{3} (I_i \omega_i)^2$, as can be verified from Eq. (7.19).

Lastly, in the absence of torque ($\mathbf{N} = 0$), the force-free Euler equations are

$$\left. \begin{array}{c} I_1 \dot{\omega}_1 - \omega_2 \omega_3 (I_2 - I_3) = 0 \\ I_2 \dot{\omega}_2 - \omega_3 \omega_1 (I_3 - I_1) = 0 \\ I_3 \dot{\omega}_3 - \omega_1 \omega_2 (I_1 - I_2) = 0 \end{array} \right\}. \qquad (7.21)$$

It is quite clear that, while these equations possess the Lagrangian

$$L = \frac{1}{2} \boldsymbol{\omega} \cdot \mathbf{I} \cdot \boldsymbol{\omega}, \qquad (7.22)$$

the force-free Euler equations (7.21) cannot be obtain from a variational principle $\delta \int L \, dt = 0$ with arbitrary variations $\delta \boldsymbol{\omega}$. Instead, by using the constrained variation[1]

$$\delta \boldsymbol{\omega} \equiv \dot{\boldsymbol{\xi}} + \boldsymbol{\omega} \times \boldsymbol{\xi}, \qquad (7.23)$$

where $\boldsymbol{\xi}$ is an arbitrary vector that vanishes at the end points (a dot refers to a time derivative in the rotating frame), we find

$$\delta L = \delta \boldsymbol{\omega} \cdot \mathbf{L} = \left(\dot{\boldsymbol{\xi}} + \boldsymbol{\omega} \times \boldsymbol{\xi} \right) \cdot \mathbf{L}$$

$$= \frac{d}{dt} \left(\boldsymbol{\xi} \cdot \mathbf{L} \right) - \boldsymbol{\xi} \cdot \left(\frac{d\mathbf{L}}{dt} + \boldsymbol{\omega} \times \mathbf{L} \right).$$

When this expression is now inserted in the variational principle $\delta \int L \, dt = 0$, we now readily obtain $d\mathbf{L}/dt + \boldsymbol{\omega} \times \mathbf{L} = 0$, from which Eqs. (7.21) are obtained.

7.2.2 Euler Equations for a Force-Free Symmetric Top

As an application of the Euler equations (7.19) we consider the case of the dynamics of a force-free symmetric top, for which $\mathbf{N} = 0$ and $I_1 = I_2 \neq I_3$. Accordingly, the Euler equations (7.19) become

$$\left. \begin{array}{c} I_1 \dot{\omega}_1 = \omega_2 \omega_3 (I_1 - I_3) \\ I_1 \dot{\omega}_2 = \omega_3 \omega_1 (I_3 - I_1) \\ I_3 \dot{\omega}_3 = 0 \end{array} \right\}, \qquad (7.24)$$

---

[1]D. D. Holm, J. E. Marsden, and T. S. Ratiu, *The EulerPoincaré Equations and Semidirect Products with Applications to Continuum Theories*, Advances in Mathematics **137**, 1-81 (1998).

The last Euler equation states that if $I_3 \neq 0$, we have $\dot{\omega}_3 = 0$, i.e., ω_3 is a constant of motion. Next, after defining the precession frequency

$$\omega_p = \omega_3 \left(\frac{I_3}{I_1} - 1 \right), \tag{7.25}$$

which may be positive $(I_3 > I_1)$ or negative $(I_3 < I_1)$, the first two Euler equations yield

$$\dot{\omega}_1(t) = -\omega_p \, \omega_2(t) \quad \text{and} \quad \dot{\omega}_2(t) = \omega_p \, \omega_1(t). \tag{7.26}$$

The general solutions for $\omega_1(t)$ and $\omega_2(t)$ are

$$\omega_1(t) = \omega_0 \, \cos(\omega_p t + \phi_0) \quad \text{and} \quad \omega_2(t) = \omega_0 \, \sin(\omega_p t + \phi_0), \tag{7.27}$$

where ω_0 is a constant and ϕ_0 is an initial phase associated with initial conditions for $\omega_1(t)$ and $\omega_2(t)$. Since ω_3 and $\omega_0^2 = \omega_1^2(t) + \omega_2^2(t)$ are constant, then the magnitude of the angular velocity $\boldsymbol{\omega}$,

$$\omega = \sqrt{\omega_1^2 + \omega_2^2 + \omega_3^2},$$

is also a constant. Thus the angle α between $\boldsymbol{\omega}$ and $\hat{\mathbf{e}}_3$ is constant, with

$$\omega_3 = \omega \, \cos\alpha \quad \text{and} \quad \sqrt{\omega_1^2 + \omega_2^2} = \omega_0 = \omega \, \sin\alpha.$$

Since the magnitude of $\boldsymbol{\omega}$ is also constant, the $\boldsymbol{\omega}$-dynamics simply involves a constant rotation with frequency ω_3 and a precession motion of $\boldsymbol{\omega}$ about the $\hat{\mathbf{e}}_3$-axis with a precession frequency ω_p; as a result of precession, the vector $\boldsymbol{\omega}$ spans the *body* cone with $\omega_p > 0$ if $I_3 > I_1$ (for a pancake-shaped or oblate symmetric top) or $\omega_p < 0$ if $I_3 < I_1$ (for a cigar-shaped or prolate symmetric top).

For example, to a good approximation, Earth is an oblate spheroid (i.e., it is flattened at the poles) with

$$I_1 = \frac{1}{5} M \left(a^2 + c^2 \right) = I_2 \quad \text{and} \quad I_3 = \frac{2}{5} M a^2 > I_1,$$

where $2c = 12,714$ km is the Pole-to-Pole distance and $2a = 12,756$ km is the equatorial diameter, so that

$$\frac{I_3}{I_1} - 1 = \frac{a^2 - c^2}{a^2 + c^2} = 0.003298... = \epsilon.$$

The precession frequency (7.25) of the rotation axis of Earth is, therefore, $\omega_p = \epsilon \, \omega_3$, where $\omega_3 = 2\pi$ rad/day is the rotation frequency of the Earth. The precession motion repeats itself every ϵ^{-1} days or 303 days; the actual period is 430 days and the difference is partially due to the non-rigidity of Earth and the fact that the Earth is not a pure oblate spheroid. A

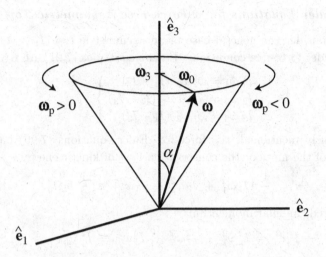

Fig. 7.3   Body cone in the Body frame of a force-free symmetric top.

slower precession motion of approximately 26,000 years is introduced by the combined gravitational effect of the Sun and the Moon on one hand, and the fact that the Earth's rotation axis is at an angle $23.5°$ to the Ecliptic plane (near which most planets move).

The fact that the symmetric top is force-free implies that its rotational kinetic energy is constant [see Eq. (7.20)] and, hence, $\mathbf{L} \cdot \boldsymbol{\omega}$ is constant while $\boldsymbol{\omega} \times \mathbf{L} \cdot \hat{\mathbf{e}}_3 = 0$ according to Eq. (7.18). Since $\mathbf{L}$ itself is constant in magnitude and direction in the LAB (or fixed) frame, we may choose the $\hat{\mathbf{z}}$-axis to be along $\mathbf{L}$ (i.e., $\mathbf{L} = \ell \hat{\mathbf{z}}$). If at a given instant, $\omega_1 = 0$, then $\omega_2 = \omega_0 = \omega \sin \alpha$ and $\omega_3 = \omega \cos \alpha$. Likewise, we may write $L_1 = I_1 \omega_1 = 0$, and

$$L_2 = I_2 \omega_2 = I_1 \omega \sin \alpha = \ell \sin \theta,$$
$$L_3 = I_3 \omega_3 = I_3 \omega \cos \alpha = \ell \cos \theta,$$

where $\mathbf{L} \cdot \boldsymbol{\omega} = \ell \omega \cos \theta$, with θ represents the *space*-cone angle. From these equations, we find the relation between the body-cone angle α and the space-cone angle θ to be

$$\tan \theta = \left(\frac{I_1}{I_3} \right) \tan \alpha, \tag{7.28}$$

which shows that $\theta > \alpha$ for $I_3 < I_1$ and $\theta < \alpha$ for $I_3 > I_1$.

7.2.3   *Euler Equations for a Force-Free Asymmetric Top*

We now consider the general case of an asymmetric top $(I_1 > I_2 > I_3)$ moving under force-free conditions. Euler's equations (7.21) are written as

$$\left.\begin{array}{l} I_1\,\dot{\omega}_1 = \omega_2\,\omega_3\,(I_2 - I_3) \\ I_2\,\dot{\omega}_2 = -\,\omega_3\,\omega_1\,(I_1 - I_3) \\ I_3\,\dot{\omega}_3 = \omega_1\,\omega_2\,(I_1 - I_2) \end{array}\right\}. \tag{7.29}$$

As previously mentioned, the force-free Euler equations (7.29) have two constants of the motion: the conservation laws of kinetic energy

$$\kappa \;=\; \frac{1}{2}\,\left(I_1\,\omega_1^2 \,+\, I_2\,\omega_2^2 \,+\, I_3\,\omega_3^2\right) \;\equiv\; \frac{1}{2}\,I_0\,\Omega_0^2, \tag{7.30}$$

and (squared) angular momentum

$$\ell^2 \;=\; I_1^2\,\omega_1^2 \,+\, I_2^2\,\omega_2^2 \,+\, I_3^2\,\omega_3^2 \;\equiv\; I_0^2\,\Omega_0^2, \tag{7.31}$$

which are expressed in terms of the parameters

$$I_0 \;\equiv\; \ell^2/(2\,\kappa) \quad \text{and} \quad \Omega_0 \equiv 2\,\kappa/\ell. \tag{7.32}$$

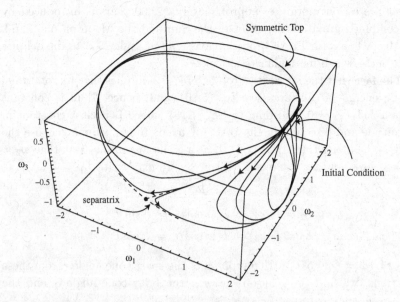

Fig. 7.4   Orbits of an asymmetric top with initial condition $(\omega_{10}, \omega_{20}, \omega_{30}) = (2, 0, 1)$ for different values of the ratio $I_1/I_3 > 1$ for a fixed ratio $I_2/I_3 > 1$.

Figure 7.4 shows the numerical solution of the Euler equations (7.29) subject to the initial condition $(\omega_{10}, \omega_{20}, \omega_{30}) = (2, 0, 1)$ for different values of the ratio $I_1/I_3 > 1$ for a fixed ratio $I_2/I_3 > 1$. Note that in the limit $I_1 = I_2$ (corresponding to a symmetric top), the top evolves solely on the (ω_1, ω_2)-plane at constant ω_3. As I_1 increases from I_2, the asymmetric top exhibits doubly-periodic behavior in the full $(\omega_1, \omega_2, \omega_3)$-space until the motion becomes restricted to the (ω_2, ω_3)-plane in the limit $I_1 \gg I_2$. One also clearly notes in Fig. 7.4 the existence of a separatrix which appears as I_1 reaches the critical value

$$I_{1c} = \frac{I_2}{2} + \sqrt{\frac{I_2^2}{4} + I_3 (I_2 - I_3) \left(\frac{\omega_{30}}{\omega_{10}}\right)^2},$$

at constant I_2 and I_3 and given initial conditions $(\omega_{10}, \omega_{20}, \omega_{30})$. This critical value is obtained by substituting $I_0 = I_2$ in Eqs. (7.30)-(7.31) and solving for I_1.

We note that the existence of two constants of the motion, Eqs. (7.30) and (7.31), for the three Euler equations (7.29) means that we may express the Euler equations in terms of a single equation. These conservation laws can be used to introduce the following definitions

$$\omega_1(\tau) = -\sqrt{\frac{I_0 (I_0 - I_3)}{I_1 (I_1 - I_3)}} \, \Omega_0 \sqrt{1 - y^2(\tau)}$$

$$\equiv -\Omega_1(I_0) \sqrt{1 - y^2(\tau)}, \tag{7.33}$$

$$\omega_2(\tau) = \sqrt{\frac{I_0 (I_0 - I_3)}{I_2 (I_2 - I_3)}} \, \Omega_0 \, y(\tau) \equiv \Omega_2(I_0) \, y(\tau), \tag{7.34}$$

$$\omega_3(\tau) = \sqrt{\frac{I_0 (I_1 - I_0)}{I_3 (I_1 - I_3)}} \, \Omega_0 \sqrt{1 - m \, y^2(\tau)}$$

$$\equiv \Omega_3(I_0) \sqrt{1 - m \, y^2(\tau)}, \tag{7.35}$$

where $\tau = [(I_1 - I_3) \Omega_1 \Omega_3 / (I_2 \Omega_2)] t$ is the dimensionless time and the modulus m in Eq. (7.35) is defined as

$$m(I_0) \equiv \frac{(I_0 - I_3) (I_1 - I_2)}{(I_2 - I_3) (I_1 - I_0)}. \tag{7.36}$$

By requiring that the modulus m be positive, the parameter I_0 introduced in Eqs. (7.30)-(7.31) must satisfy $I_3 < I_0 < I_1$ and, hence, $0 \le m(I_0) \le 1$ for $I_3 \le I_0 \le I_2$ and $m(I_0) > 1$ for $I_2 < I_0 < I_1$ (with $m \to \infty$ as $I_0 \to I_1$).

The solutions for $\omega_1(\tau)$, $\omega_2(\tau)$ and $\omega_3(\tau)$, subject to the initial conditions $(\omega_{10}, \omega_{20}, \omega_{30}) = (-\Omega_1, 0, \Omega_3)$, can be expressed in terms of the Jacobi elliptic functions (sn, cn, dn) as [12]

$$(\omega_1, \, \omega_2, \, \omega_3) = (-\Omega_1 \, \text{cn} \, \tau, \, \Omega_2 \, \text{sn} \, \tau, \, \Omega_3 \, \text{dn} \, \tau). \tag{7.37}$$

Lastly, we note that the separatrix solution of the force-free asymmetric top (see Fig. 7.4) corresponds to $I_0 = I_2$ (see problem 3).

7.3   Lagrangian Rigid-Body Dynamics

7.3.1   *Eulerian Angles as Generalized Coordinates*

The Lagrangian description of the physical state of a rotating object with principal moments of inertia (I_1, I_2, I_3) requires the definition of three Eulerian angles (φ, θ, ψ) in the body frame of reference (see Fig. 7.5).

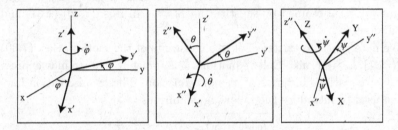

Fig. 7.5   Euler angles (φ, θ, ψ).

The first Eulerian angle φ (left figure in Fig. 7.5) is associated with the rotation of the fixed-frame unit vectors $(\hat{x}, \hat{y}, \hat{z})$ about the z-axis. Through this rotation, we thus obtain the new unit vectors $(\hat{x}', \hat{y}', \hat{z}')$ defined as

$$\begin{pmatrix} \hat{x}' \\ \hat{y}' \\ \hat{z}' \end{pmatrix} = \overbrace{\begin{pmatrix} \cos\varphi & \sin\varphi & 0 \\ -\sin\varphi & \cos\varphi & 0 \\ 0 & 0 & 1 \end{pmatrix}}^{= \, \mathsf{R}_3(\varphi)} \cdot \begin{pmatrix} \hat{x} \\ \hat{y} \\ \hat{z} \end{pmatrix} \tag{7.38}$$

The second Eulerian angle θ (center figure in Fig. 7.5) is associated with the rotation of the unit vectors $(\hat{x}', \hat{y}', \hat{z}')$ about the x'-axis. We thus obtain the new unit vectors $(\hat{x}'', \hat{y}'', \hat{z}'')$ defined as

$$\begin{pmatrix} \hat{x}'' \\ \hat{y}'' \\ \hat{z}'' \end{pmatrix} = \overbrace{\begin{pmatrix} 1 & 0 & 0 \\ 0 & \cos\theta & \sin\theta \\ 0 & -\sin\theta & \cos\theta \end{pmatrix}}^{= \, \mathsf{R}_1(\theta)} \cdot \begin{pmatrix} \hat{x}' \\ \hat{y}' \\ \hat{z}' \end{pmatrix} \tag{7.39}$$

The third Eulerian angle ψ (right figure in Fig. 7.5) is associated with the rotation of the unit vectors $(\hat{x}'', \hat{y}'', \hat{z}'')$ about the z''-axis. We finally obtain the body-frame unit vectors $(\hat{e}_1, \hat{e}_2, \hat{e}_3)$ defined as

$$
\begin{pmatrix} \hat{e}_1 \\ \hat{e}_2 \\ \hat{e}_3 \end{pmatrix} = \overbrace{\begin{pmatrix} \cos\psi & \sin\psi & 0 \\ -\sin\psi & \cos\psi & 0 \\ 0 & 0 & 1 \end{pmatrix}}^{= \, \mathsf{R}_3(\psi)} \cdot \begin{pmatrix} \hat{x}'' \\ \hat{y}'' \\ \hat{z}'' \end{pmatrix} \tag{7.40}
$$

Hence, the relation between the fixed-frame unit vectors $\hat{x}^j = (\hat{x}, \hat{y}, \hat{z})$ and the body-frame unit vectors $\hat{e}_i = (\hat{e}_1, \hat{e}_2, \hat{e}_3)$ involves the matrix $\mathsf{R} = \mathsf{R}_3(\psi) \cdot \mathsf{R}_1(\theta) \cdot \mathsf{R}_3(\varphi)$, such that $\hat{e}_i = R_{ij}\,\hat{x}^j$, or

$$
\left.\begin{aligned}
\hat{e}_1 &= \cos\psi\,\hat{\perp} + \sin\psi\,(\cos\theta\,\hat{\varphi} + \sin\theta\,\hat{z}) \\[2mm]
\hat{e}_2 &= -\sin\psi\,\hat{\perp} + \cos\psi\,(\cos\theta\,\hat{\varphi} + \sin\theta\,\hat{z}) \\[2mm]
\hat{e}_3 &= -\sin\theta\,\hat{\varphi} + \cos\theta\,\hat{z}
\end{aligned}\right\}, \tag{7.41}
$$

where $\hat{\varphi} = -\sin\varphi\,\hat{x} + \cos\varphi\,\hat{y}$ and $\hat{\perp} \equiv \hat{\varphi} \times \hat{z}$.

7.3.2 *Angular Velocity in Terms of Eulerian Angles*

According to Fig. 7.5, the angular velocity ω is expressed in terms of the frequencies $(\dot{\varphi}, \dot{\theta}, \dot{\psi})$ as

$$
\omega = \dot{\varphi}\,\hat{z} + \dot{\theta}\,\hat{x}' + \dot{\psi}\,\hat{e}_3. \tag{7.42}
$$

The unit vectors $\hat{z}$ and $\hat{x}'$ are written in terms of the body-frame unit vectors $(\hat{e}_1, \hat{e}_2, \hat{e}_3)$ as

$$
\begin{aligned}
\hat{z} &= \sin\theta\,(\sin\psi\,\hat{e}_1 + \cos\psi\,\hat{e}_2) + \cos\theta\,\hat{e}_3, \\
\hat{x}' &= \hat{x}'' = \cos\psi\,\hat{e}_1 - \sin\psi\,\hat{e}_2.
\end{aligned}
$$

The angular velocity (7.42) can, therefore, be written exclusively in the body frame of reference in terms of the Euler basis vectors (7.41) as

$$
\omega = \omega_1\,\hat{e}_1 + \omega_2\,\hat{e}_2 + \omega_3\,\hat{e}_3, \tag{7.43}
$$

where the body-frame angular frequencies are

$$
\left.\begin{aligned}
\omega_1 &= \dot{\varphi}\,\sin\theta\,\sin\psi + \dot{\theta}\,\cos\psi \\
\omega_2 &= \dot{\varphi}\,\sin\theta\,\cos\psi - \dot{\theta}\,\sin\psi \\
\omega_3 &= \dot{\psi} + \dot{\varphi}\,\cos\theta
\end{aligned}\right\}. \tag{7.44}
$$

Note that all three frequencies are independent of φ (i.e., $\partial\omega_i/\partial\varphi = 0$), while derivatives with respect to ψ and $\dot{\psi}$ are

$$
\frac{\partial\omega_1}{\partial\psi} = \omega_2, \quad \frac{\partial\omega_2}{\partial\psi} = -\omega_1, \quad \text{and} \quad \frac{\partial\omega_3}{\partial\psi} = 0,
$$

and

$$\frac{\partial \omega_1}{\partial \dot{\psi}} = 0 = \frac{\partial \omega_2}{\partial \dot{\psi}} \quad \text{and} \quad \frac{\partial \omega_3}{\partial \dot{\psi}} = 1.$$

The relations (7.44) can be inverted to yield

$$\left.\begin{array}{l} \dot{\varphi} = \csc\theta \, (\sin\psi \, \omega_1 + \cos\psi \, \omega_2) \\ \dot{\theta} = \cos\psi \, \omega_1 - \sin\psi \, \omega_2 \\ \dot{\psi} = \omega_3 - \cot\theta \, (\sin\psi \, \omega_1 + \cos\psi \, \omega_2) \end{array}\right\}. \tag{7.45}$$

Equations (7.44) allow us to relate Lagrangian rigid-body dynamics, expressed in terms of the angles (φ, θ, ψ) and their derivatives $(\dot{\varphi}, \dot{\theta}, \dot{\psi})$, to Eulerian rigid-body dynamics expressed in terms of the angular velocity $\boldsymbol{\omega} = \omega_1 \widehat{e}_1 + \omega_2 \widehat{e}_2 + \omega_3 \widehat{e}_3$.

7.3.3 *Rotational Kinetic Energy of a Symmetric Top*

The rotational kinetic energy (7.5) for a symmetric top can be written as

$$K_{rot} = \frac{1}{2} \left[I_3 \, \omega_3^2 + I_1 \left(\omega_1^2 + \omega_2^2 \right) \right],$$

or explicitly in terms of the Eulerian angles (φ, θ, ψ) and their time derivatives $(\dot{\varphi}, \dot{\theta}, \dot{\psi})$ as

$$K_{rot} = \frac{1}{2} \left[I_3 \left(\dot{\psi} + \dot{\varphi} \, \cos\theta \right)^2 + I_1 \left(\dot{\theta}^2 + \dot{\varphi}^2 \, \sin^2\theta \right) \right]. \tag{7.46}$$

We now briefly return to the case of the force-free symmetric top for which the Lagrangian is simply $L(\theta, \dot{\theta}; \dot{\varphi}, \dot{\psi}) = K_{rot}$.

Since the Eulerian angles φ and ψ are ignorable coordinates, i.e., the force-free Lagrangian (7.46) is independent of φ and ψ, their canonical angular momenta

$$p_\varphi = \frac{\partial L}{\partial \dot{\varphi}} = I_3 \left(\dot{\psi} + \dot{\varphi} \, \cos\theta \right) \cos\theta + I_1 \sin^2\theta \, \dot{\varphi}, \tag{7.47}$$

$$p_\psi = \frac{\partial L}{\partial \dot{\psi}} = I_3 \left(\dot{\psi} + \dot{\varphi} \, \cos\theta \right) \equiv I_3 \, \omega_3 \tag{7.48}$$

are constants of the motion. By inverting these relations, we obtain

$$\dot{\varphi} = \frac{p_\varphi - p_\psi \, \cos\theta}{I_1 \sin^2\theta} \quad \text{and} \quad \dot{\psi} = \omega_3 - \frac{(p_\varphi - p_\psi \, \cos\theta) \, \cos\theta}{I_1 \sin^2\theta}, \tag{7.49}$$

and the rotational kinetic energy (7.46) becomes

$$K_{rot} = \frac{1}{2} \left[I_1 \, \dot{\theta}^2 + I_3 \, \omega_3^2 + \frac{(p_\varphi - p_\psi \, \cos\theta)^2}{I_1 \sin^2\theta} \right]. \tag{7.50}$$

The Routhian function for the free symmetric top (with Lagrangian $L = K_{rot}$) is therefore

$$
\begin{aligned}
R(\theta, \dot{\theta}; p_\varphi, p_\psi) &= L - p_\varphi \dot{\varphi} - p_\psi \dot{\psi} \\
&= \frac{I_1}{2} \dot{\theta}^2 - \frac{(p_\varphi - p_\psi \cos\theta)^2}{2\, I_1 \sin^2\theta},
\end{aligned} \tag{7.51}
$$

where the second term represents the effective potential for the force-free symmetric top and the constant term $- I_3 \omega_3^2 / 2$ was removed.

The motion of a force-free symmetric top can now be described in terms of solutions of the Euler-Lagrange equation for the Eulerian angle θ:

$$
\begin{aligned}
\frac{d}{dt} \left(\frac{\partial R}{\partial \dot{\theta}} \right) &= I_1 \ddot{\theta} = \frac{\partial R}{\partial \theta} = \dot{\varphi} \sin\theta \, (I_1 \cos\theta \, \dot{\varphi} - p_\psi) \\
&= - \frac{(p_\varphi - p_\psi \cos\theta)}{I_1 \sin\theta} \frac{(p_\psi - p_\varphi \cos\theta)}{\sin^2\theta}.
\end{aligned} \tag{7.52}
$$

Once $\theta(t)$ is solved for given values of the principal moments of inertia $I_1 = I_2$ and I_3 and the invariant canonical angular momenta p_φ and p_ψ, the functions $\varphi(t)$ and $\psi(t)$ are determined from the time integration of Eqs. (7.49).

7.3.4 *Symmetric Top with One Fixed Point*

We now consider the case of a spinning symmetric top of mass M and principal moments of inertia $(I_1 = I_2 \neq I_3)$ with one fixed point O moving in a gravitational field with constant acceleration g (see Fig. 7.6). The rotational kinetic energy of the symmetric top is given by Eq. (7.46) while the potential energy for the case of a symmetric top with one fixed point is

$$
U(\theta) = Mgh \cos\theta, \tag{7.53}
$$

where h is the distance from the fixed point O and the center of mass (CM) of the symmetric top.

The Routhian for the symmetric top with one fixed point (also known as the *heavy* symmetric top) is

$$
R(\theta, \dot{\theta}; p_\varphi, p_\psi) = \frac{1}{2} I_1 \dot{\theta}^2 - \left[\frac{1}{2} \frac{(p_\varphi - p_\psi \cos\theta)^2}{I_1 \sin^2\theta} + Mgh \cos\theta \right]. \tag{7.54}
$$

A normalized form of the Euler equations for the symmetric top with one fixed point is expressed as

$$
\varphi' = \frac{(b - \cos\theta)}{\sin^2\theta} \quad \text{and} \quad \theta'' = a \sin\theta - \frac{(1 - b\cos\theta)(b - \cos\theta)}{\sin^3\theta}, \tag{7.55}
$$

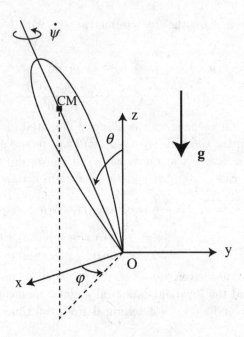

Fig. 7.6   Symmetric top with one fixed point.

where time has been rescaled such that $(\cdots)' = (I_1/p_\psi)\,d(\cdots)/dt$ and the two parameters a and b are defined as $a = Mgh\,I_1/p_\psi^2$ and $b = p_\varphi/p_\psi$.

We note that the azimuthal equation of motion φ' can change direction if $b - \cos\theta$ changes sign. The normalized heavy-top equations (7.55) have been integrated for the initial conditions $(\theta_0, \theta_0';\ \varphi_0) = (1, 0; 0)$. The three Figures shown below correspond to three different cases (I, II, and III) for fixed value of a (here, $a = 0.1$), which exhibit the possibility of azimuthal reversal when φ' changes sign for different values of $b = p_\varphi/p_\psi$; the azimuthal precession motion is called *nutation*.

Figures 7.7-7.9 show (on the left) the normalized heavy-top solutions in the (φ, θ)-plane ($\cos\theta$ increases downward) and (on the right) the spherical projection of the normalized heavy-top solutions $(\theta, \varphi) \rightarrow (\sin\theta\,\cos\varphi,\ \sin\theta\,\sin\varphi,\ \cos\theta)$, where the initial condition is denoted by a dot ($\bullet$). In Case I ($b > \cos\theta_0$), the azimuthal velocity φ' never changes sign and azimuthal precession occurs monotonically. In Case II ($b = \cos\theta_0$), the azimuthal velocity φ' vanishes at $\theta = \theta_0$ (where θ' also vanishes) and the heavy symmetric top exhibits a *cusp* at $\theta = \theta_0$. In Case III ($b < \cos\theta_0$), the azimuthal velocity φ' vanishes for $\theta > \theta_0$ and the heavy symmetric top

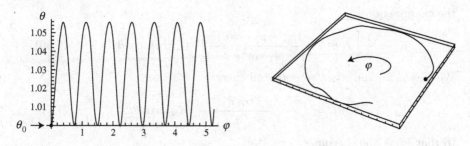

Fig. 7.7 Orbits of a heavy top – Case I ($b > \cos\theta_0 > \cos\theta$).

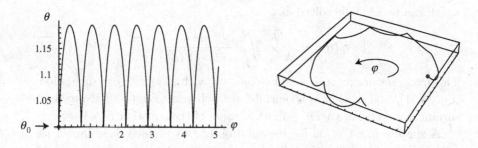

Fig. 7.8 Orbits of a heavy top – Case II ($b = \cos\theta_0 > \cos\theta$).

exhibits a phase of *retrograde* motion.

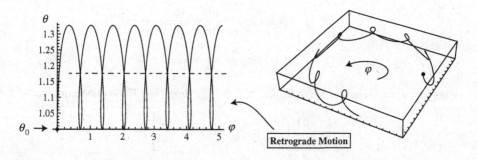

Fig. 7.9 Orbits of a heavy top – Case III ($\cos\theta_0 > b > \cos\theta$).

Since the Lagrangian (7.54) is independent of the Eulerian angles φ and ψ, the canonical angular momenta p_φ and p_ψ, respectively, are constants of the motion. The solution for $\theta(t)$ is then most easily obtained by considering the energy equation

$$E = \frac{1}{2}\left[I_1\dot{\theta}^2 + \frac{(p_\varphi - p_\psi \cos\theta)^2}{I_1 \sin^2\theta}\right] + Mgh\cos\theta. \qquad (7.56)$$

We may define an effective potential energy

$$V(\theta) = \frac{(p_\varphi - p_\psi \cos\theta)^2}{2I_1 \sin^2\theta} + Mgh\cos\theta, \qquad (7.57)$$

so that Eq. (7.56) becomes

$$E = \frac{1}{2}I_1\dot{\theta}^2 + V(\theta), \qquad (7.58)$$

which can be formally solved as

$$t(\theta) = \sqrt{\frac{I_1}{2}}\int_{\theta_0}^{\theta}\frac{d\phi}{\sqrt{E - V(\phi)}}. \qquad (7.59)$$

This integral solution is in complete analogy with motion in one dimension [Eq. (3.31)] and motion in a central-force potential [Eq. (4.9)]. Note that turning points θ_{tp} are again defined as roots of the equation $E = V(\theta)$.

A simpler formulation for the solution of this problem is obtained as follows. First, we define the following quantities

$$\Omega^2 = \frac{2Mgh}{I_1}, \quad \epsilon = \frac{2E}{I_1\Omega^2} = \frac{E}{Mgh}, \quad \alpha = \frac{p_\varphi}{I_1\Omega}, \text{ and } \beta = \frac{p_\psi}{I_1\Omega}, \tag{7.60}$$

so that Eq. (7.59) becomes

$$\tau(u) = \pm\int\frac{du}{\sqrt{(1-u^2)(\epsilon - u) - (\alpha - \beta u)^2}}$$

$$= \pm\int\frac{du}{\sqrt{(1-u^2)[\epsilon - W(u)]}}, \qquad (7.61)$$

where $\tau(u) = \Omega t(u)$, $u = \cos\theta$, and the energy equation (7.58) becomes

$$\epsilon = \frac{1}{(1-u^2)}\left[\left(\frac{du}{d\tau}\right)^2 + (\alpha - \beta u)^2\right] + u$$

$$= (1-u^2)^{-1}\left(\frac{du}{d\tau}\right)^2 + W(u), \qquad (7.62)$$

where the effective potential is

$$W(u) = u + \frac{(\alpha - \beta u)^2}{(1-u^2)}.$$

We note that the effective potential $W(u)$ is infinite at $u = \pm 1$ and has a single minimum at $u = u_0$ (or $\theta = \theta_0$) defined by the quartic equation

$$W'(u_0) = 1 + 2u_0 \left(\frac{\alpha - \beta u_0}{1 - u_0^2}\right)^2 - 2\beta \left(\frac{\alpha - \beta u_0}{1 - u_0^2}\right) = 0. \qquad (7.63)$$

This equation has four roots: two roots are complex roots, a third root is always greater than one for $\alpha > 0$ and $\beta > 0$ (which is unphysical since $u = \cos\theta \leq 1$), while the fourth root is less than one for $\alpha > 0$ and $\beta > 0$; hence, this root is the only physical root corresponding to a single minimum for the effective potential $W(u)$ (see Fig. 7.10). Note how the linear gravitational-potential term u is apparent at low values of α and β.

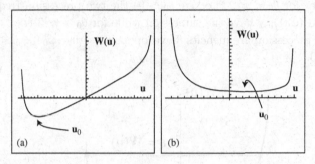

Fig. 7.10   Effective potential $W(u)$ for the heavy top. (a) Minimum at $\theta_0 > \pi/2$ ($u_0 < 0$) for $\alpha = 0.1$ and $\beta = 0.5$ and (b) minimum at $\theta_0 < \pi/2$ ($u_0 > 0$) for $\alpha = 2.0$ and $\beta = 1.0$.

We first investigate the motion of the symmetric top at the minimum angle θ_0 for which $\epsilon = W(u_0)$ and $\dot{u}_0 = 0$. For this purpose, we note that when the dimensionless azimuthal frequency

$$\frac{d\varphi}{d\tau} = \frac{\alpha - \beta u}{1 - u^2} = \nu(u)$$

is inserted in Eq. (7.63), we obtain the quadratic equation for $\nu_0 = \nu(u_0)$:

$$1 + 2u_0 \nu_0^2 - 2\beta \nu_0 = 0,$$

which has two solutions

$$\nu_0 = \frac{\beta}{2u_0} \left(1 \pm \sqrt{1 - \frac{2u_0}{\beta^2}}\right). \qquad (7.64)$$

Here, we further note that these solutions require that the radicand in Eq. (7.64) be positive, i.e.,

$$\beta^2 > 2u_0 \quad \text{or} \quad I_3 \omega_3 \geq I_1 \Omega \sqrt{2u_0},$$

if $u_0 \geq 0$ (or $\theta_0 \leq \pi/2$); no condition is applied to ω_3 for the case $u_0 < 0$ (or $\theta_0 > \pi/2$) since the radicand is strictly positive in this case.

Hence, the precession frequency $\dot{\varphi}_0 = \nu(u_0)\,\Omega$ at $\theta = \theta_0$ has a slow component and a fast component

$$(\dot{\varphi}_0)_{slow} = \frac{I_3\omega_3}{2\,I_1\cos\theta_0}\left[1 - \sqrt{1 - 2\left(\frac{I_1\Omega}{I_3\omega_3}\right)^2\cos\theta_0}\,\right],$$

$$(\dot{\varphi}_0)_{fast} = \frac{I_3\omega_3}{2\,I_1\cos\theta_0}\left[1 + \sqrt{1 - 2\left(\frac{I_1\Omega}{I_3\omega_3}\right)^2\cos\theta_0}\,\right].$$

We note that for $\theta_0 < \pi/2$ (i.e., $\cos\theta_0 > 0$) the two precession frequencies $(\dot{\varphi}_0)_{slow}$ and $(\dot{\varphi}_0)_{fast}$ have the same sign while for $\theta_0 > \pi/2$ (i.e., $\cos\theta_0 < 0$) the two precession frequencies have opposite signs $(\dot{\varphi}_0)_{slow} < 0$ and $(\dot{\varphi}_0)_{fast} > 0$.

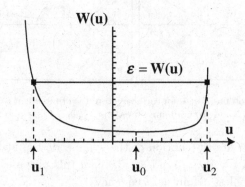

Fig. 7.11   Turning-point roots.

Next, we investigate the case with two turning points $u_1 < u_0 < u_2$ (or $\theta_1 > \theta_2$) (see Fig. 7.11), where the θ-dynamics oscillates between θ_1 and $\theta_2 < \theta_1$. The turning points u_1 and u_2 are roots of the cubic polynomial

$$F(u) = (1 - u^2)\,[\epsilon - W(u)]$$
$$= u^3 - (\epsilon + \beta^2)\,u^2 - (1 - 2\alpha\beta)\,u + (\epsilon - \alpha^2). \qquad (7.65)$$

Although a third root u_3 exists for $F(u) = 0$, it is unphysical since $u_3 > 1$. Since the azimuthal frequencies at the turning points are expressed as

$$\frac{d\varphi_1}{d\tau} = \frac{\alpha - \beta\,u_1}{1 - u_1^2} \quad \text{and} \quad \frac{d\varphi_2}{d\tau} = \frac{\alpha - \beta\,u_2}{1 - u_2^2},$$

where $\alpha - \beta u_1 > \alpha - \beta u_2$, we can study the three cases for nutation numerically investigated (see Figures 7.7-7.9); here, we assume that both $\alpha = b\beta$ and β are positive. In Case I ($\alpha > \beta u_2$), the precession frequency $d\varphi/d\tau$ is strictly positive for $u_1 \le u \le u_2$ and nutation proceeds monotonically. In Case II ($\alpha = \beta u_2$), the precession frequency $d\varphi/d\tau$ is positive for $u_1 \le u < u_2$ and vanishes at $u = u_2$; nutation in this Case exhibits a cusp at θ_2. In Case III ($\alpha < \beta u_2$), the precession frequency $d\varphi/d\tau$ reverses its sign at $u_r = \alpha/\beta = b$ or $\theta_2 < \theta_r = \arccos(b) < \theta_1$.

7.3.5 *Stability of the Sleeping Top*

Let us consider the case where a symmetric top with one fixed point is launched with initial conditions $\theta_0 \ne 0$ and $\dot\theta_0 = \dot\varphi_0 = 0$, with $\dot\psi_0 \ne 0$. In this case, the invariant canonical momenta are $p_\psi = I_3 \dot\psi_0$ and $p_\varphi = p_\psi \cos\theta_0$ (i.e., the initial conditions correspond to Case II). These initial conditions ($u_0 = \alpha/\beta, \dot u_0 = 0$), therefore, imply from Eq. (7.62) that $\epsilon = u_0$ and that the energy equation (7.62) becomes

$$\left(\frac{du}{d\tau}\right)^2 = \left[(1 - u^2) - \beta^2 (u_0 - u) \right] (u_0 - u). \qquad (7.66)$$

We consider the case of the *sleeping* top with the additional initial condition $\theta_0 = 0$ (and $u_0 = 1$). Thus Eq. (7.66) becomes

$$\left(\frac{du}{d\tau}\right)^2 = (1 + u - \beta^2) (1 - u)^2. \qquad (7.67)$$

The sleeping top has the following turning points (where $u' = 0$): $\overline{u}_1 = 1$ (a double root of the cubic polynomial) and $\overline{u}_2 = \beta^2 - 1$. The sleeping top also has equilibrium points (where $u'' = 0$) at $u_1 = 1$ and $u_2 = (2\beta^2 - 1)/3$. Equation (7.67) can be written in the Weierstrass elliptic form [see Eq. (B.19)] for $w \equiv (u - \alpha)/4$, where $\alpha = (\beta^2 + 1)/3$ and the Weierstrass invariants are $g_2 = 12(\alpha - 1)^2$ and $g_3 = 2(\alpha - 1)^3$. The equilibrium points are written in terms of w as $w_1 = (2 - \beta^2)/12 = -w_2$. The solution of the sleeping top can therefore be entirely expressed in terms of the Weierstrass elliptic function (which could be done as an exercise); see Appendix B for details on elliptic functions. Instead of pursuing this solution, however, we note that it is quite clear that the case $\beta^2 = 2$ presents an important transition since the two equilibrium points merge at $w = 0$.

We now investigate the stability of the equilibrium point $u_1 = 1$ by writing $u = 1 - \delta$ (with $\delta \ll 1$) so that Eq. (7.67) becomes

$$\frac{d\delta}{d\tau} = (2 - \beta^2)^{\frac{1}{2}} \delta.$$

The solution of this equation is exponential (and, therefore, u_1 is unstable) if $\beta^2 < 2$ or oscillatory (and, therefore, u_1 is stable) if $\beta^2 > 2$. Note that in the latter case, the condition $\beta^2 > 2$ implies that the second equilibrium point $u_2 > 1$ is unphysical. The stability of the sleeping top therefore requires a large spinning frequency ω_3 (so that $\beta^2 > 2$); in the presence of friction, the spinning frequency slows down and ultimately the sleeping top becomes unstable. Bifurcation theory deals with the investigation of the stability equilibrium points as a parameter is varied, which is one of the topics studied in the next Chapter.

7.4 Problems

1. Consider a thin homogeneous rectangular plate of mass M and area $a\,b$ that lies on the (x, y)-plane.

(a) Show that the inertia tensor (calculated in the reference frame with its origin at one corner of the plate) takes the form

$$\mathbf{I} = \begin{pmatrix} A & -C & 0 \\ -C & B & 0 \\ 0 & 0 & A+B \end{pmatrix},$$

and find suitable expressions for A, B, and C in terms of M, a, and b.

(b) Show that by performing a rotation of the coordinate axes about the z-axis through an angle θ, the new inertia tensor is

$$\mathbf{I}'(\theta) = \mathsf{R}(\theta) \cdot \mathbf{I} \cdot \mathsf{R}^T(\theta) = \begin{pmatrix} A' & -C' & 0 \\ -C' & B' & 0 \\ 0 & 0 & A'+B' \end{pmatrix},$$

where

$$A' = A \cos^2 \theta + B \sin^2 \theta - C \sin 2\theta$$
$$B' = A \sin^2 \theta + B \cos^2 \theta + C \sin 2\theta$$
$$C' = C \cos 2\theta - \frac{1}{2}(B - A) \sin 2\theta.$$

(c) When

$$\theta = \frac{1}{2} \arctan\left(\frac{2C}{B - A}\right),$$

the off-diagonal component C' vanishes and the $x'-$ and $y'-$axes become principal axes. Calculate expressions for A' and B' in terms of M, a, and b for this particular angle.

(d) Calculate the inertia tensor $\mathbf{I}_{CM}$ in the CM frame by using the Parallel-Axis Theorem and show that

$$I_{CM}^x = \frac{M b^2}{12}, \quad I_{CM}^y = \frac{M a^2}{12}, \quad \text{and} \quad I_{CM}^z = \frac{M}{12}\left(b^2 + a^2\right).$$

2. Derive the moment of inertia (7.16).

3. (a) The Euler equation for an asymmetric top $(I_1 > I_2 > I_3)$ with $L^2 = 2\,I_2 K$ is $\dot{\omega}_2 = \alpha\left(\Omega^2 - \omega_2^2\right)$, where

$$\Omega^2 = \frac{2K}{I_2} \quad \text{and} \quad \alpha = \sqrt{\left(1 - \frac{I_2}{I_1}\right)\left(\frac{I_2}{I_3} - 1\right)}$$

Solve for $\omega_2(t)$ with the initial condition $\omega_2(0) = 0$.

(b) Use the solution $\omega_2(t)$ found in Part (a) to find the solutions $\omega_1(t)$ and $\omega_3(t)$ given by Eqs. (7.33) and (7.35) for $m = 1$.

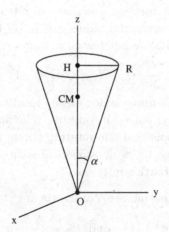

Fig. 7.12 Problem 4.

4. Consider a circular cone of height H and base radius $R = H \tan\alpha$ with uniform mass density $\rho = 3 M/(\pi H R^2)$.

(a) Show that the non-vanishing components of the inertia tensor **I** calculated from the vertex O of the cone are

$$I_{xx} = I_{yy} = \frac{3}{5} M \left(H^2 + \frac{R^2}{4} \right) \quad \text{and} \quad I_{zz} = \frac{3}{10} M R^2$$

(b) Show that the principal moments of inertia calculated in the CM frame (located at a height $h = 3H/4$ on the symmetry axis) are

$$I_1 = I_2 = \frac{3}{20} M \left(R^2 + \frac{H^2}{4} \right) \quad \text{and} \quad I_3 = \frac{3}{10} M R^2$$

5. Show that the Euler basis vectors $(\widehat{e}_1, \widehat{e}_2, \widehat{e}_3)$ are determined by Eq. (7.41).

6. (a) Use Euler's equations (7.19) to find the torque needed to rotate a rectangular plate of sides a and b about a diagonal with constant angular velocity $\boldsymbol{\omega}$.

(b) Show that this torque vanishes if the rectangular plate is a square (i.e., if $a = b$).

7. As a result of its daily rotation, the shape of Earth is approximated as an oblate spheroid with equatorial radius $a = 6,378$ km and polar radius $c = 6,357$ km. The gravitational potential is expressed as

$$\Phi(r, \theta) = -\frac{GM}{r} + \frac{G}{2 r^3} (I_3 - I_1) (3 \cos^2 \theta - 1) - \frac{1}{2} \omega^2 r^2 \sin^2 \theta,$$

where M denotes Earth's mass, ω denotes its rotation angular frequency, and G is the gravitational universal constant. The first term is the gravitational potential for a spherical non-rotating Earth, the second term represents the correction due to its non-spherical shape, and the third term represents the effects of Earth's rotation.

(a) Show that the principle moments of inertia are

$$I_1 = \frac{M}{5} (a^2 + c^2) \quad \text{and} \quad I_3 = \frac{2 M}{5} a^2 > I_1.$$

(b) Compute the gravitational acceleration $\mathbf{g} \equiv - \nabla \Phi$, and calculate its magnitude on the equator $(r, \theta) = (a, \pi/2)$ and at the north pole $(r, \theta) =$

$(c, 0)$.

(c) Compare the directions and magnitudes of the corrections to the gravitational acceleration due to the centrifugal term and the non-spherical term.

8. In the absence of external torque, the Euler equations (7.19) can be written as

$$\frac{dL_i}{dt} = \{L_i, K_{rot}\} = -\hat{\imath} \cdot \boldsymbol{\omega} \times \mathbf{L} = -\epsilon_{ijk}\, \omega_j \left(I_k\, \omega_k \right),$$

where the *Poisson bracket* { , } is defined in terms of two arbitrary functions $F(\mathbf{L})$ and $G(\mathbf{L})$ as

$$\{F, G\} = -\mathbf{L} \cdot \frac{\partial F}{\partial \mathbf{L}} \times \frac{\partial G}{\partial \mathbf{L}}.$$

Hence a general function $F(\mathbf{L})$ of angular momentum evolves according to the Hamilton's equation

$$\frac{dF}{dt} = \{F, K_{rot}\} = -\frac{\partial F}{\partial \mathbf{L}} \cdot \boldsymbol{\omega} \times \mathbf{L}.$$

(a) Show that any function of $|\mathbf{L}|$ is a constant of the motion for rigid body dynamics.

(b) Show that the Poisson bracket satisfies the Jacobi identity

$$\{F, \{G, H\}\} + \{G, \{H, F\}\} + \{H, \{F, G\}\} = 0,$$

for three arbitrary functions F, G, and H.

Chapter 8

Normal-Mode Analysis

8.1 Stability of Equilibrium Points

The nonlinear (one-dimensional) force equation $m\ddot{x} = f(x)$ has equilibrium points (labeled x_0) where $f(x_0)$ vanishes. The stability of the equilibrium point x_0 is determined by the sign of $f'(x_0)$: the equilibrium point x_0 is stable if $f'(x_0) < 0$ or unstable if $f'(x_0) > 0$. Since $f(x)$ is also derived from a potential $V(x)$ as $f(x) = -V'(x)$, we say that the equilibrium point x_0 is stable (or unstable) if $V''(x_0)$ is positive (or negative).

8.1.1 *Sleeping Top*

As a first example, we return to the case of the sleeping top (Sec. 7.3.5) for which the equation of motion

$$u'' = \frac{3}{2} u^2 - u \left(\beta^2 + 1\right) + \left(\beta^2 - \frac{1}{2}\right) \equiv -V'(u) \qquad (8.1)$$

is derived from Eq. (7.67). This equation has two equilibrium points at $u_1 = 1$ and $u_2 = (2\beta^2 - 1)/3$. The second-order derivatives of the effective potential $V(u)$ evaluated at these equilibrium points are

$$V''(u_1) = \beta^2 - 2 = -V''(u_2),$$

which immediately implies that the two equilibrium points exchange stability properties as β^2 crosses 2. Hence, for $\beta^2 < 2$, the equilibrium point u_1 is unstable (while u_2 is stable), while for $\beta^2 > 2$, the equilibrium point u_1 is stable (while u_2 is unstable but, since $u_2 > 1$, this root becomes unphysical).

Figure 8.1 shows the *bifurcation* diagram for the problem of the sleeping top. We see that the equilibrium point $u = 1$, which is unstable for $\beta^2 < 2$,

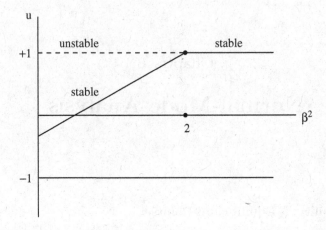

Fig. 8.1 Bifurcation diagram for the sleeping top. Note that the physical motion for $u = \cos\theta$ is only possible for $-1 \le u \le 1$.

becomes stable for $\beta^2 > 2$. The critical value $\beta^2 = 2$ is called a bifurcation point for the sleeping top.

8.1.2 *Bead on a Rotating Hoop*

As a second example, we considered in Chap. 2 the problem of a bead of mass m sliding freely on a hoop of radius R rotating with angular velocity Ω in a constant gravitational field with acceleration g. The Lagrangian for this system is

$$L(\theta, \dot\theta) = \frac{m}{2} R^2 \dot\theta^2 + \left(\frac{m}{2} R^2 \Omega^2 \sin^2\theta + mgR \cos\theta \right) = \frac{m}{2} R^2 \dot\theta^2 - V(\theta),$$

where $V(\theta)$ denotes the effective potential, and the Euler-Lagrange equation for θ is

$$mR^2 \ddot\theta = -V'(\theta) = -mR^2\Omega^2 \sin\theta \, (\nu - \cos\theta), \qquad (8.2)$$

where $\nu = g/(R\Omega^2)$. The equilibrium points of Eq. (8.2) are $\theta = 0$ (for all values of ν) and $\theta = \arccos(\nu)$ if $\nu < 1$. The stability of the equilibrium point $\theta = \theta_0$ is determined by the sign of

$$V''(\theta_0) = mR^2\Omega^2 \left[\nu \cos\theta_0 - (2\cos^2\theta_0 - 1) \right].$$

Hence,

$$V''(0) = mR^2\Omega^2 \, (\nu - 1) \qquad (8.3)$$

is positive (i.e., $\theta = 0$ is stable) if $\nu > 1$ or negative (i.e., $\theta = 0$ is unstable) if $\nu < 1$. In the latter case, when $\nu < 1$ and the second equilibrium point $\theta_0 = \arccos(\nu)$ is allowed, we find

$$V''(\theta_0) = mR^2\Omega^2 \left[\nu^2 - (2\nu^2 - 1) \right] = mR^2\Omega^2 \left(1 - \nu^2\right) > 0, \quad (8.4)$$

and thus the equilibrium point $\theta_0 = \arccos(\nu)$ is stable when $\nu < 1$.

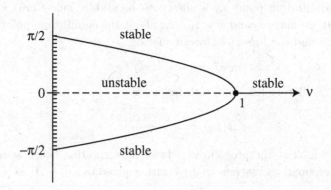

Fig. 8.2   Bifurcation diagram for the bead on a rotating-hoop problem.

Figure 8.2 shows the bifurcation diagram for the problem of a bead on a rotating hoop. Here, we see that for $\nu > 1$, a single stable equilibrium exists at $\theta = 0$. For $\nu < 1$, however, the equilibrium point $\theta = 0$ is unstable and new stable equilibrium points appear at $\theta = \pm \arccos(\nu)$. The critical value $\nu = 1$ is the bifurcation point for the bead on a rotating hoop.

8.1.3   Circular Orbits in Central-Force Fields

As our last example in this Section, we consider the radial force equation

$$\mu \ddot{r} = \frac{\ell^2}{\mu r^3} - k r^{n-1} = -V'(r),$$

studied in Chap. 4 for a central-force field $F(r) = -k r^{n-1}$ (here, μ is the reduced mass of the system, the azimuthal angular momentum ℓ is a constant of the motion, and k is a constant). The equilibrium point at $r = \rho$ is defined by the relation $V'(\rho) = 0$:

$$\rho^{n+2} = \frac{\ell^2}{\mu k}. \quad (8.5)$$

The second derivative of the effective potential is

$$V''(\rho) = \frac{\ell^2}{\mu \rho^4} \left(3 + (n-1) \frac{k\mu}{\ell^2} \rho^{n+2} \right) = \frac{\ell^2}{\mu\rho^4} (2+n). \quad (8.6)$$

Hence, $V''(\rho)$ is positive if $n > -2$, and, thus, circular orbits are stable in central-force fields $F(r) = -k\,r^{n-1}$ if $n > -2$.

8.2   Small Oscillations about Stable Equilibria

Once an equilibrium point x_0 is shown to be stable, i.e., $f'(x_0) < 0$ or $V''(x_0) > 0$, we may expand $x = x_0 + \delta x$ about the equilibrium point (with $\delta x \ll x_0$) to find the *linearized* force equation

$$m\,\delta\ddot{x} = -V''(x_0)\,\delta x, \qquad (8.7)$$

which has oscillatory behavior with frequency

$$\omega(x_0) = \sqrt{\frac{V''(x_0)}{m}}.$$

We first look at the problem of a bead on a rotating hoop, where the frequency of small oscillations $\omega(\theta_0)$ is either given in Eq. (8.3) as

$$\omega(0) = \sqrt{\frac{V''(0)}{mR^2}} = \omega_0\,\sqrt{\nu - 1}$$

for $\theta_0 = 0$ and $\nu > 1$, or is given in Eq. (8.4) as

$$\omega(\theta_0) = \sqrt{\frac{V''(\theta_0)}{mR^2}} = \omega_0\,\sqrt{1 - \nu^2}$$

for $\theta_0 = \arccos(\nu)$ and $\nu < 1$.

Next, we look at the frequency of small oscillations about the stable circular orbit in a central-force field $F(r) = -k\,r^{n-1}$ (with $n < 3$). Here, from Eq. (8.6), we find

$$\omega = \sqrt{\frac{V''(\rho)}{\mu}} = \sqrt{\frac{k\,(2+n)}{\mu\,\rho^{2-n}}},$$

where $\ell^2 = \mu k\,\rho^{2+n}$ was used. We note that for the Kepler problem $(n = -1)$, the period of small oscillations $T = 2\pi/\omega$ is expressed as

$$T^2 = \frac{(2\pi)^2\,\mu}{k}\,\rho^3,$$

which is precisely the statement of Kepler's Third Law for circular orbits [see Eq. (4.28)]. Hence, a small perturbation of a stable Keplerian circular orbit does not change its orbital period.

As a last example of linear stability, we consider the case of a time-dependent equilibrium. A rigid parabolic wire having equation $z = k\,r^2$

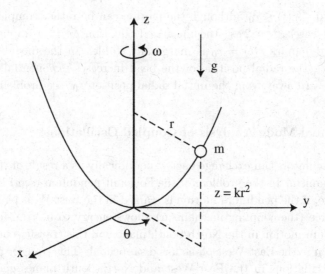

Fig. 8.3   Bead on a rotating parabolic wire.

is fastened to a vertical shaft rotating at constant angular velocity $\dot{\theta} = \omega$. A bead of mass m is free to slide along the wire in the presence of a constant gravitational field with potential $U(z) = mg\,z$ (see Fig. 8.3). The Lagrangian for this mechanical system is given as

$$L(r, \dot{r}) \;=\; \frac{m}{2}\left(1 + 4\,k^2 r^2\right)\dot{r}^2 \;+\; m\left(\frac{\omega^2}{2} - g\,k\right)r^2,$$

and the Euler-Lagrange equation of motion is easily obtained as

$$\left(1 + 4\,k^2 r^2\right)\ddot{r} \;+\; 4\,k^2 r\,\dot{r}^2 \;=\; \left(\omega^2 - 2\,gk\right)r. \tag{8.8}$$

Note that when $\omega^2 < 2\,gk$, we see that the bead moves in an effective potential represented by an isotropic simple harmonic oscillator with spring constant $\sqrt{m\left(2\,gk - \omega^2\right)}$ (i.e., the radial position of the bead is bounded), while when $\omega^2 > 2\,gk$, the bead appears to move on the surface of an inverted paraboloid and, thus, the radial position of the bead in this case is unbounded.

We now investigate the stability of the linearized motion $r(t) = r_0 + \delta r(t)$ about an initial radial position r_0. The linearized equation for $\delta r(t)$ is

$$\delta\ddot{r} \;=\; \left(\frac{\omega^2 - 2\,gk}{1 + 4\,k^2 r_0^2}\right)\delta r,$$

so that the radial position $r = r_0$ is stable if $\omega^2 < 2\,gk$ and unstable if $\omega^2 > 2\,gk$. In the stable case ($\omega^2 < 2\,gk$), the bead oscillates back and forth with $0 \leq r(t) \leq r_0$, although the motion can be rather complex. For the special case $\omega^2 = 2\,gk$, the linearized equation $\delta\ddot{r} = 0$ implies that the radial dynamics $r(t) = r_0$ is marginally stable. In the unstable case $(\omega^2 > 2\,gk)$, the radial position of the bead increases exponentially as it spirals outward away from the initial radial position r_0 (see problem 1).

8.3 Normal-Mode Analysis of Coupled Oscillations

Coupled oscillators can exchange energy periodically as a result of the coupling mechanism. In the problem of the Foucault pendulum (see Fig. 6.6), for example, if the pendulum motion is started in the East-West plane, the Coriolis force (the coupling mechanism) allows energy to be transfered to the pendulum motion in the North-South plane and this transfer continues until motion in the East-West plane has disappeared. This transfer process between oscillations in the East-West and North-South planes generates the standard precession motion of the Foucault pendulum.

The normal-mode analysis enables us to determine the characteristic oscillation frequencies exhibited by coupled linear oscillators. For nonlinear coupled oscillators, the nonlinear equations of motion must be linearized first before obtaining the characteristic frequencies of small oscillations.

8.3.1 *Coupled Simple Harmonic Oscillators*

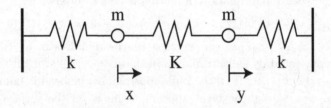

Fig. 8.4 Coupled identical masses and springs.

We begin our study of linearly-coupled oscillators by considering the following coupled system comprised of two block-and-spring systems, with identical mass m and identical spring constant k, coupled by means of a spring of constant K (see Fig. 8.4). The coupled equations are

$$m\ddot{x} = -(k+K)\,x + K\,y \quad \text{and} \quad m\ddot{y} = -(k+K)\,y + K\,x. \quad (8.9)$$

The solutions for $x(t)$ and $y(t)$ are obtained by using a method known as the normal-mode analysis. First, we write $x(t)$ and $y(t)$ in the normal-mode representation

$$x(t) = \overline{x}\, e^{-i\omega t} \quad \text{and} \quad y(t) = \overline{y}\, e^{-i\omega t}, \tag{8.10}$$

where $\overline{x}$ and $\overline{y}$ are constant oscillation amplitudes and the eigenfrequency ω is to be solved in terms of the system parameters (m, k, K). Next, substituting the normal-mode representation (8.10) into Eq. (8.9), we obtain the following normal-mode matrix equation

$$\begin{pmatrix} \omega^2 m - (k+K) & K \\ K & \omega^2 m - (k+K) \end{pmatrix} \begin{pmatrix} \overline{x} \\ \overline{y} \end{pmatrix} = 0, \tag{8.11}$$

which couples the amplitudes $\overline{x}$ and $\overline{y}$. To obtain a non-trivial solution $(\overline{x}, \overline{y}) \neq (0, 0)$, the determinant of the matrix in Eq. (8.11) is required to vanish, which yields the characteristic polynomial

$$[\omega^2 m - (k+K)]^2 - K^2 = 0,$$

whose solutions are the eigenfrequencies

$$\omega_{\pm}^2 = \frac{(k+K)}{m} \pm \frac{K}{m}.$$

If we insert $\omega_{+}^2 = (k + 2K)/m$ into the matrix equation (8.11), we obtain

$$\begin{pmatrix} K & K \\ K & K \end{pmatrix} \begin{pmatrix} \overline{x} \\ \overline{y} \end{pmatrix} = 0,$$

which implies that $\overline{y} = -\overline{x}$, and thus the *eigenfrequency* ω_{+} is associated with an *antisymmetric* coupled motion. This antisymmetric coupled motion implies that the position of the center-of-mass for the two identical particles remains constant. If we insert $\omega_{-}^2 = k/m$ into the matrix equation (8.11), on the other hand, we obtain

$$\begin{pmatrix} -K & K \\ K & -K \end{pmatrix} \begin{pmatrix} \overline{x} \\ \overline{y} \end{pmatrix} = 0,$$

which implies that $\overline{y} = \overline{x}$, and thus the eigenfrequency ω_{-} is associated with a *symmetric* coupled motion. For this symmetric motion, the position of the center-of-mass oscillates with frequency ω_{-}.

Figure 8.5 shows the normalized solution of the coupled equations (8.9), where time is normalized as $t \to \sqrt{k/m}\, t$ for the weak coupling $(K < k)$ case. Note that the two eigenfrequencies are said to be *commensurate* if the ratio $\omega_{+}/\omega_{-} = \sqrt{1 + 2K/k}$ is expressed as a rational number ρ for values of the ratio $K/k = (\rho^2 - 1)/2$ and that for commensurate eigenfrequencies, the

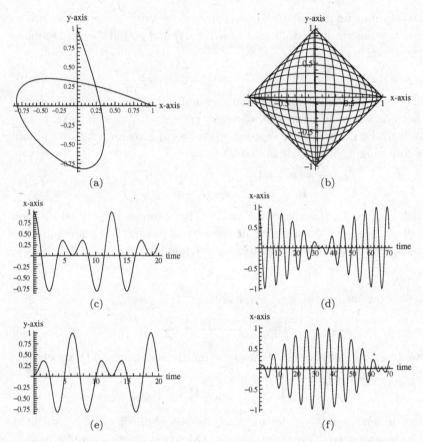

Fig. 8.5   Weak-coupling normalized solutions of the coupled equations (8.9) for $K/k = 5/8$ (plots a, c, e) and $K/k = 0.1$ (plots b, d, f). Plots (a)-(b) show the parametric plots of $y(t)$ versus $x(t)$; plots (c)-(d) show $x(t)$ versus time t; and plots (e)-(f) show $x(t)$ versus time t.

graph of the solutions on the (x, y)-plane generates the so-called Lissajous figures. For non-commensurate eigenfrequencies, however, the graph of the solutions on the (x, y)-plane shows more complex behavior.

Lastly, we construct the normal coordinates η_+ and η_-, which satisfy the condition $\ddot{\eta}_\pm = -\omega_\pm^2\, \eta_\pm$. From the discussion above, we find

$$\eta_-(t) = x(t) + y(t) \quad \text{and} \quad \eta_+(t) = x(t) - y(t). \tag{8.12}$$

Figure 8.6 shows the graphs of the normal coordinates $\eta_\pm(t) = x(t) \mp y(t)$, which clearly displays the single-frequency behavior predicted by the present normal-mode analysis. The solutions $\eta_\pm(t)$ are of the form

$$\eta_\pm = A_\pm\, \cos(\omega_\pm t + \varphi_\pm),$$

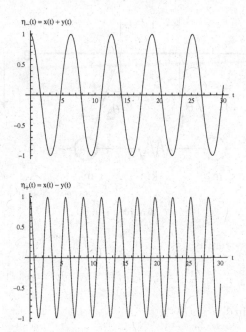

Fig. 8.6   Normal coordinates $\eta_-(t)$ and $\eta_+(t)$ as a function of normalized time for the case $K/k = 2$ with normalized frequencies 1 and $\sqrt{5}$, respectively.

where $A_\pm$ and $\varphi_\pm$ are constants (determined from initial conditions). The general solution of Eqs. (8.9) can, therefore, be written explicitly in terms of the normal coordinates $\eta_\pm$ as

$$\begin{pmatrix} x(t) \\ y(t) \end{pmatrix} = \frac{A_-}{2} \cos(\omega_- t + \varphi_-) \pm \frac{A_+}{2} \cos(\omega_+ t + \varphi_+).$$

8.3.2   *Coupled Nonlinear Oscillators*

We now consider the following system composed of two pendula of identical length ℓ but different masses m_1 and m_2 coupled by means of a spring of constant k in the presence of a gravitational field of constant acceleration g (see Fig. 8.7). Here, the distance D between the two points of attach of the pendula is equal to the length of the spring in its relaxed state and we assume, for simplicity, that the masses always stay on the same horizontal line.

Using the generalized coordinates (θ_1, θ_2) defined in Fig. 8.7, the La-

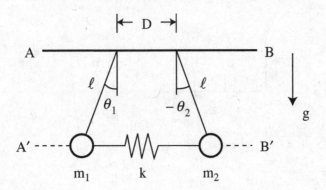

Fig. 8.7   Coupled pendula.

grangian for this system is

$$L = \frac{\ell^2}{2} \left(m_1\, \dot\theta_1^2 + m_2\, \dot\theta_2^2 \right) - g\ell \left[m_1 \left(1 - \cos\theta_1 \right) + m_2 \left(1 - \cos\theta_2 \right) \right]$$
$$- \frac{k\ell^2}{2} \left(\sin\theta_1 - \sin\theta_2 \right)^2,$$

and the nonlinear coupled equations of motion are

$$\left. \begin{array}{l} m_1\, \ddot\theta_1 = - m_1\, \omega_g^2\, \sin\theta_1 - k \left(\sin\theta_1 - \sin\theta_2 \right) \cos\theta_1 \\[2mm] m_2\, \ddot\theta_2 = - m_2\, \omega_g^2\, \sin\theta_2 + k \left(\sin\theta_1 - \sin\theta_2 \right) \cos\theta_2 \end{array} \right\}, \qquad (8.13)$$

where $\omega_g^2 = g/\ell$.

It is quite clear that the equilibrium point is $\theta_1 = 0 = \theta_2$ and expansion of the coupled equations (8.13) about this equilibrium yields the coupled linear equations

$$\left. \begin{array}{l} m_1\, \ddot{q}_1 = - m_1\, \omega_g^2\, q_1 - k \left(q_1 - q_2 \right) \\[2mm] m_2\, \ddot{q}_2 = - m_2\, \omega_g^2\, q_2 + k \left(q_1 - q_2 \right), \end{array} \right\}, \qquad (8.14)$$

where $\theta_1 = q_1 \ll 1$ and $\theta_2 = q_2 \ll 1$. The normal-mode matrix associated with these coupled linear equations (8.14) is

$$\begin{pmatrix} (\omega^2 - \omega_g^2)\, m_1 - k & k \\ k & (\omega^2 - \omega_g^2)\, m_2 - k \end{pmatrix} \begin{pmatrix} q_1 \\ q_2 \end{pmatrix} = 0, \qquad (8.15)$$

and the characteristic polynomial is

$$M \left[(\omega^2 - \omega_g^2)\, \mu - k \right] (\omega^2 - \omega_g^2) = 0,$$

where $\mu = m_1 m_2 / M$ is the reduced mass for the system and $M = m_1 + m_2$ is the total mass. The eigenfrequencies are thus

$$\omega_-^2 = \omega_g^2 \quad \text{and} \quad \omega_+^2 = \omega_g^2 + \frac{k}{\mu}.$$

The normal coordinates $\eta_\pm$ are expressed in terms of (q_1, q_2) as $\eta_\pm = a_\pm q_1 + b_\pm q_2$, where $a_\pm$ and $b_\pm$ are constant coefficients determined from the condition $\ddot{\eta}_\pm = -\omega_\pm^2 \eta_\pm$. By inserting the eigenfrequency $\omega_-^2 = \omega_g^2$ into Eq. (8.15), we find $b_-/a_- = 1$ (i.e., masses move symmetrically), and thus we may choose

$$\eta_- = q_1 + q_2.$$

By inserting the eigenfrequency $\omega_+^2 = \omega_g + k/\mu$ into Eq. (8.15), on the other hand, we find $b_+/a_+ = -m_1/m_2$, and thus we may choose

$$\eta_+ = \frac{m_1}{M} q_1 - \frac{m_2}{M} q_2,$$

which represents the center of mass for the system. Lastly, we may solve for q_1 and q_2 as

$$q_1 = \eta_+ + \frac{m_2}{M} \eta_- \quad \text{and} \quad q_2 = -\eta_+ + \frac{m_1}{M} \eta_+,$$

where $\eta_\pm = A_\pm \cos(\omega_\pm t + \varphi_\pm)$ are general solutions of the normal-mode equations $\ddot{\eta}_\pm = -\omega_\pm^2 \eta_\pm$.

8.4 Problems

1. This problem deals with the numerical integration of the radial equation of motion (8.8), which is expressed in dimensionless form as

$$(1 + 4\rho^2)\,\rho'' + 4\rho\,(\rho')^2 = (\Omega - 2)\,\rho, \qquad (8.16)$$

where Ω is a dimensionless parameter and $\rho' = d\rho/d\tau$ is defined in terms of the dimensionless time τ.

(a) Find expressions for ρ, τ, and Ω.

(b) Integrate Eq. (8.16) for (I) $\Omega < 2$, (II) $\Omega = 2$, and (III) $\Omega > 2$.

(c) Compare the orbits obtained in Part (b) with the stability analysis of this problem found in Sec. 8.2.

2. The following compound pendulum is composed of two identical masses m attached by massless rods of identical length ℓ to a ring of mass M, which

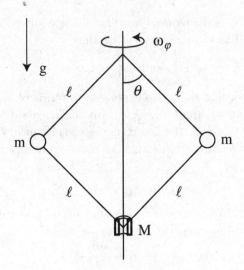

Fig. 8.8   Problem 2.

is allowed to slide up and down along a vertical axis in a gravitational field with constant g (see Fig. 8.8). The entire system rotates about the vertical axis with an azimuthal angular frequency ω_φ.

(a) Show that the Lagrangian for the system can be written as

$$L(\theta, \dot\theta) \;=\; \ell^2 \dot\theta^2 \left(m + 2M \sin^2\theta\right) \;+\; m\ell^2\, \omega_\varphi^2 \sin^2\theta \;+\; 2\left(m + M\right)g\ell\, \cos\theta$$

(b) Identify the equilibrium points for the system and investigate their stability.

(c) Determine the frequency of small oscillations about each stable equilibrium point found in Part (b).

3. Consider the same problem as in Sec. (8.3.1) but now with different masses $m_1 \neq m_2$ (see Fig. 8.9). Calculate the eigenfrequencies and eigenvectors (normal coordinates) for this system.

4. Find the eigenfrequencies associated with small oscillations of the system shown in Fig. 8.10.

5. Two blocks of identical mass m are attached by massless springs (with identical spring constant k) as shown in Fig. 8.11. The Lagrangian for this

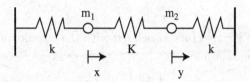

Fig. 8.9 Problem 3.

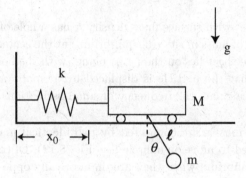

Fig. 8.10 Problem 4.

system is

$$L(x, \dot{x}; y, \dot{y}) = \frac{m}{2}\left(\dot{x}^2 + \dot{y}^2\right) - \frac{k}{2}\left[x^2 + (y - x)^2\right],$$

where x and y denote departures from equilibrium.

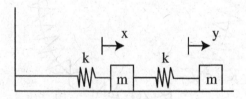

Fig. 8.11 Problem 5.

(a) Derive the Euler-Lagrange equations for x and y.

(b) Show that the eigenfrequencies for small oscillations for this system are

$$\omega_{\pm}^2 = \frac{\omega_k^2}{2}\left(3 \pm \sqrt{5}\right),$$

where $\omega_k^2 = k/m$.

(c) Show that the eigenvectors associated with the eigenfrequencies $\omega_\pm$ are represented by the relations

$$\overline{y}_\pm = \frac{1}{2}\left(1 \mp \sqrt{5}\right)\overline{x}_\pm$$

where $(\overline{x}_\pm,\ \overline{y}_\pm)$ represent the normal-mode amplitudes.

6. An infinite sheet with surface mass density σ has a hole of radius R cut into it. A particle of mass m sits (in equilibrium) at the center of the circle. Assuming that the sheet lies on the (x, y)-plane (with the hole centered at the origin) and that the particle is displaced by a small amount $z \ll R$ along the z-axis, calculate the frequency of small oscillations.

7. Two identical masses are connected by two identical massless springs and are constrained to move on a circle (see Fig. 8.12). Of course, the two masses are in equilibrium when they are diametrically opposite points on the circle. Solve for the normal modes of the system.

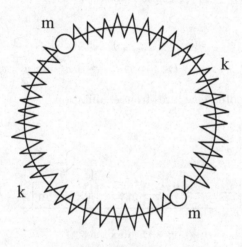

Fig. 8.12   Problem 7.

8. Consider a pendulum of mass m attached at a point O with the help of a massless rigid rod on length ℓ. Here, point O is located at a distance $R > \ell$ from a axis of rotation and is rotating at an angular velocity Ω about

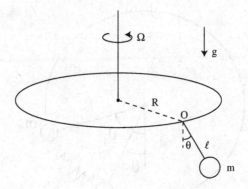

Fig. 8.13   Problem 8.

the axis of rotation (see Fig. 8.13),

(a) Show that there are two equilibrium configurations to this problem, which are obtained from finding the roots to the transcendental equation

$$(R - \ell \sin \theta) \, \Omega^2 \cos \theta = g \sin \theta.$$

(b) Show that one equilibrium configuration is stable while the other is unstable.

9. Two particles of identical masses are connected to each other by a spring (with constant k) and are allowed to move without friction on a hoop of radius R (see Fig. 8.14). The angles θ_1 and θ_2 are expressed in terms of the generalized coordinates θ and φ as $\theta_1 = \theta$ and $\theta_2 = \theta + \varphi - \Theta$, where θ is the angular displacement of the right mass from the vertical, Θ is the angular separation between the two masses when the spring is at equilibrium, and φ is the angular displacement of the spring away from equilibrium.

(a) Show that the Lagrangian for this system is

$$L = \frac{m}{2} R^2 \left[\dot\theta^2 + \left(\dot\theta + \dot\varphi \right)^2 \right] - \frac{k}{2} (R\varphi)^2 + mgR \left[\cos \theta + \cos (\theta + \varphi - \Theta) \right],$$

and derive the Euler-Lagrange equations for θ and φ.

(b) Show that the potential $U(\theta, \varphi)$ has a global minimum at θ_0 and φ_0, which satisfy the transcendental equation

$$\theta_0 = \frac{1}{2} (\Theta - \varphi_0) \equiv \frac{1}{2} \left(\Theta - \Omega^2 \sin \theta_0 \right),$$

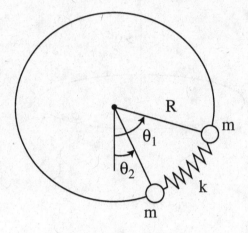

Fig. 8.14   Problem 9.

where $\Omega^2 \equiv \omega_g^2/\omega_k^2$ is the ratio of the squared pendulum frequency $\omega_g^2 = g/R$ and the squared spring frequency $\omega_k^2 = k/m$.

(c) Find the eigenfrequencies and eigenvectors for the normal modes of small oscillations about the equilibrium defined by θ_0 and φ_0.

Chapter 9

Continuous Lagrangian Systems

This last Chapter represents in fact the beginning for some of the most important applications of Lagrangian methods in physics, namely those that apply to classical, general relativistic, or quantum field theories. So far, Lagrangian methods have been applied to derive equations of motion for particles or rigid bodies. The Noether method has also been applied to obtain the conservation laws of energy and momentum for these systems whenever symmetries existed for the corresponding Lagrangians.

While a systematic presentation of the Lagrangian formulation of field equations cannot be undertaken at this level, a few examples have nonetheless been selected that give a flavor of the power of the Lagrangian method for continuous systems.

9.1 Waves on a Stretched String

9.1.1 *Wave Equation*

The equation describing transverse waves propagating on a stretched string of constant linear mass density ρ under constant tension T is

$$\rho \, \frac{\partial^2 u(x,t)}{\partial t^2} \; = \; T \, \frac{\partial^2 u(x,t)}{\partial x^2}, \tag{9.1}$$

where $u(x,t)$ denotes the amplitude of the wave at position x along the string at time t. General solutions to this linear wave equation involve arbitrary functions $g(x \pm v\,t)$, where $v = \sqrt{T/\rho}$ represents the speed of waves propagating on the string. Indeed, we find

$$\rho \, \partial_t^2 g(x \pm v\,t) \; = \; \rho v^2 \, g'' \; = \; T \, g'' \; = \; T \, \partial_x^2 g(x \pm v\,t).$$

The interpretation of the two different signs is that $g(x - v\,t)$ represents a wave propagating to the right while $g(x+v\,t)$ represents a wave propagating to the left. The general solution of the wave equation (9.1) is

$$u(x, t) = A_- \, g(x - v\,t) + A_+ \, g(x + v\,t),$$

where $A_\pm$ are arbitrary constants determined from initial conditions

9.1.2 *Lagrangian Formulation*

The question we now ask is whether the wave equation (9.1) can be derived from a variational principle

$$\delta \int \mathcal{L}(u, \partial_t u, \partial_x u) \, dx \, dt = 0, \tag{9.2}$$

where the Lagrangian *density* $\mathcal{L}(u, \partial_t u, \partial_x u)$ is a function of the dynamical variable $u(x, t)$ and its space-time derivatives. Here, the variation of the Lagrangian density $\mathcal{L}$ is expressed as

$$\delta\mathcal{L} = \delta u \, \frac{\partial \mathcal{L}}{\partial u} + \partial_t \delta u \, \frac{\partial \mathcal{L}}{\partial(\partial_t u)} + \partial_x \delta u \, \frac{\partial \mathcal{L}}{\partial(\partial_x u)},$$

where $\delta u(x, t)$ is a general variation of $u(x, t)$ subject to the condition that it vanishes at the integration boundaries in Eq. (9.2). By re-arranging terms, the variation of $\mathcal{L}$ can be written as

$$\delta\mathcal{L} = \delta u \left\{ \frac{\partial \mathcal{L}}{\partial u} - \frac{\partial}{\partial t}\left(\frac{\partial \mathcal{L}}{\partial(\partial_t u)} \right) - \frac{\partial}{\partial x}\left(\frac{\partial \mathcal{L}}{\partial(\partial_x u)} \right) \right\}$$
$$+ \frac{\partial}{\partial t}\left(\delta u \, \frac{\partial \mathcal{L}}{\partial(\partial_t u)} \right) + \frac{\partial}{\partial x}\left(\delta u \, \frac{\partial \mathcal{L}}{\partial(\partial_x u)} \right). \tag{9.3}$$

When we insert this expression for $\delta\mathcal{L}$ into the variational principle (9.2), we obtain

$$\int dx \, dt \, \delta u \left\{ \frac{\partial \mathcal{L}}{\partial u} - \frac{\partial}{\partial t}\left(\frac{\partial \mathcal{L}}{\partial(\partial_t u)} \right) - \frac{\partial}{\partial x}\left(\frac{\partial \mathcal{L}}{\partial(\partial_x u)} \right) \right\} = 0, \tag{9.4}$$

where the last two terms in Eq. (9.3) cancel out because δu vanishes on the integration boundaries. Since the variational principle (9.4) is true for general variations δu, we obtain the Euler-Lagrange equation for the dynamical field $u(x, t)$:

$$\frac{\partial}{\partial t}\left(\frac{\partial \mathcal{L}}{\partial(\partial_t u)} \right) + \frac{\partial}{\partial x}\left(\frac{\partial \mathcal{L}}{\partial(\partial_x u)} \right) = \frac{\partial \mathcal{L}}{\partial u}. \tag{9.5}$$

The question we posed earlier now focuses on deciding what form the Lagrangian density must take. Here, the answer is surprisingly simple: the kinetic energy density of the wave is $\rho \, (\partial_t u)^2 / 2$, while the potential energy density is $T\,(\partial_x u)^2/2$, and thus the Lagrangian density for waves on a stretched string is

$$\mathcal{L}(u, \partial_t u, \partial_x u; \; x, t) = \frac{\rho}{2}\left(\frac{\partial u}{\partial t}\right)^2 - \frac{T}{2}\left(\frac{\partial u}{\partial x}\right)^2. \qquad (9.6)$$

Since $\partial\mathcal{L}/\partial u = 0$, we find

$$\frac{\partial}{\partial t}\left(\frac{\partial\mathcal{L}}{\partial(\partial_t u)}\right) = \frac{\partial}{\partial t}\left(\rho\,\frac{\partial u}{\partial t}\right) = \rho\,\frac{\partial^2 u}{\partial t^2},$$

$$\frac{\partial}{\partial x}\left(\frac{\partial\mathcal{L}}{\partial(\partial_x u)}\right) = \frac{\partial}{\partial x}\left(-T\,\frac{\partial u}{\partial x}\right) = -T\,\frac{\partial^2 u}{\partial x^2},$$

and Eq. (9.1) is indeed represented as an Euler-Lagrange equation (9.5) in terms of the Lagrangian density (9.6).

The energy density $\mathcal{E}$ of a stretched string can also be calculated by using the Legendre transformation:

$$\mathcal{E} \equiv \frac{\partial u}{\partial t}\,\frac{\partial\mathcal{L}}{\partial(\partial_t u)} - \mathcal{L} = \frac{\rho}{2}\left(\frac{\partial u}{\partial t}\right)^2 + \frac{T}{2}\left(\frac{\partial u}{\partial x}\right)^2.$$

By using the wave equation (9.1), we readily find that the time derivative of the energy density

$$\frac{\partial\mathcal{E}}{\partial t} = \frac{\partial}{\partial x}\left(T\,\frac{\partial u}{\partial t}\,\frac{\partial u}{\partial x}\right)$$

can be expressed as an energy conservation law $\partial_t\mathcal{E} + \partial\mathcal{S} = 0$, where the energy-density flux is defined as $\mathcal{S} \equiv -T\,\partial_x u\,\partial_t u$. The next Section will present the general variational formulation of classical field theory, which enables us to show that the wave equation (9.1) also satisfies the momentum conservation law $\partial_t\mathcal{P} + \partial_x\mathcal{E} = 0$, where the momentum density is $\mathcal{P} \equiv \mathcal{S}/v^2$.

9.2 Variational Principle for Field Theory*

The simple example of transverse waves on a stretched string allows us to view the Euler-Lagrange equation (9.5) as a generalization of the Euler-Lagrange equations

$$\frac{d}{dt}\left(\frac{\partial L}{\partial\dot{q}^i}\right) = \frac{\partial L}{\partial q^i},$$

in terms of the generalized coordinates q^i. We now spend some time investigating the Lagrangian description of continuous systems, in which the dynamical variable is the single-component field $\psi(\mathbf{x}, t)$, which can easily be generalized to fields with multiple components.

9.2.1 *Lagrangian Formulation*

Classical and quantum field theories rely on variational principles based on the existence of action functionals. The typical action functional is of the form

$$\mathcal{A}[\psi] \;=\; \int d^4x\; \mathcal{L}(\psi,\, \partial_\mu\psi), \tag{9.7}$$

where the wave function $\psi(\mathbf{x}, t)$ represents the state of the system at position $\mathbf{x}$ (in n-dimensional space) and time t while the entire physical content of the theory is carried by the Lagrangian density $\mathcal{L}$. In addition, we use the convenient four-vector notation $\partial_\mu = (c^{-1}\partial_t,\, \nabla)$ in Eq. (9.7) and we use the space-like metric tensor[1] $g^{\mu\nu} = \mathrm{diag}(-1, +1, +1, +1)$.

The variational principle is based on the stationarity of the action functional

$$0 \;=\; \delta\mathcal{A}[\psi] \;=\; \frac{d}{d\epsilon}\left(\mathcal{A}[\psi + \epsilon\,\delta\psi]\right)_{\epsilon=0} \;=\; \int \delta\mathcal{L}(\psi,\, \partial_\mu\psi)\; d^4x. \tag{9.8}$$

Here, the functional variation of the Lagrangian density is

$$\delta\mathcal{L} = \frac{\partial\mathcal{L}}{\partial\psi}\,\delta\psi \;+\; \frac{\partial\mathcal{L}}{\partial(\partial_\mu\psi)}\,\partial_\mu\delta\psi$$

$$\equiv \delta\psi\left[\frac{\partial\mathcal{L}}{\partial\psi} - \frac{\partial}{\partial x^\mu}\left(\frac{\partial\mathcal{L}}{\partial(\partial_\mu\psi)}\right)\right] + \frac{\partial\Lambda^\mu}{\partial x^\mu}, \tag{9.9}$$

where

$$\frac{\partial\mathcal{L}}{\partial(\partial_\mu\psi)}\,\partial_\mu\delta\psi = \frac{\partial\mathcal{L}}{\partial(\nabla\psi)}\cdot\nabla\delta\psi + \frac{\partial\mathcal{L}}{\partial(\partial_t\psi)}\,\partial_t\delta\psi,$$

and the exact space-time divergence $\partial_\mu\Lambda^\mu$ in Eq. (9.9) is obtained by rearranging terms, with

$$\Lambda^\mu = \delta\psi\,\frac{\partial\mathcal{L}}{\partial(\partial_\mu\psi)} \quad\text{and}\quad \frac{\partial\Lambda^\mu}{\partial x^\mu} = \frac{\partial}{\partial t}\left(\delta\psi\,\frac{\partial\mathcal{L}}{\partial(\partial_t\psi)}\right) + \nabla\cdot\left(\delta\psi\,\frac{\partial\mathcal{L}}{\partial(\nabla\psi)}\right).$$

The variational principle (9.8) then yields

$$0 = \int d^4x\; \delta\psi\left[\frac{\partial\mathcal{L}}{\partial\psi} - \frac{\partial}{\partial x^\mu}\left(\frac{\partial\mathcal{L}}{\partial(\partial_\mu\psi)}\right)\right],$$

where the exact divergence $\partial_\mu\Lambda^\mu$ drops out under the assumption that the variation $\delta\psi$ vanish on the integration boundaries. Following the standard rules of Calculus of Variations, the Euler-Lagrange equation for the field ψ is

$$\frac{\partial}{\partial x^\mu}\left(\frac{\partial\mathcal{L}}{\partial(\partial_\mu\psi)}\right) = \frac{\partial}{\partial t}\left(\frac{\partial\mathcal{L}}{\partial(\partial_t\psi)}\right) + \nabla\cdot\left(\frac{\partial\mathcal{L}}{\partial(\nabla\psi)}\right) = \frac{\partial\mathcal{L}}{\partial\psi}. \tag{9.10}$$

The generalization to a multiple-component field simply involves replacing the single field ψ with the field component ψ^a (where the component index $a \geq 2$).

---

[1]For two four-vectors $A^\mu = (A^0, \mathbf{A})$ and $B^\mu = (B^0, \mathbf{B})$, we have $A \cdot B = A_\mu B^\mu = \mathbf{A}\cdot\mathbf{B} - A^0 B^0$, where $A_0 = -A^0$.

9.2.2 Noether Method and Conservation Laws

Since the Euler-Lagrange equation (9.10) holds true for arbitrary field variations $\delta\psi$, the variation of the Lagrangian density $\mathcal{L}$ is now expressed as the Noether equation

$$\delta\mathcal{L} \equiv \partial_\mu \Lambda^\mu = \frac{\partial}{\partial x^\mu}\left[\delta\psi\,\frac{\partial\mathcal{L}}{\partial(\partial_\mu\psi)}\right], \qquad (9.11)$$

which associates symmetries with conservation laws $\partial_\mu\,\mathcal{J}^\mu = 0$.

9.2.2.1 Energy-Momentum Conservation Law

The conservation of energy-momentum (a four-vector quantity) involves a symmetry of the Lagrangian with respect to constant *space-time* translations $(x^\nu \to x^\nu + \delta x^\nu)$. The variation $\delta\psi$ is no longer arbitrary in Eq. (9.9) but is required to be of the form

$$\delta\psi = -\,\delta x^\nu\,\partial_\nu\psi \qquad (9.12)$$

while the variation $\delta\mathcal{L}$ is of the form

$$\delta\mathcal{L} = -\,\delta x^\nu\left[\partial_\nu\mathcal{L} - (\partial_\nu\mathcal{L})_\psi\right], \qquad (9.13)$$

where $(\partial_\nu\mathcal{L})_\psi$ denotes a explicit space-time derivative of $\mathcal{L}$ at constant ψ. The Noether equation (9.11) can now be written as

$$\frac{\partial}{\partial x^\mu}\left(\mathcal{L}\,g^\mu{}_\nu - \frac{\partial\mathcal{L}}{\partial(\partial_\mu\psi)}\,\partial_\nu\psi\right) = \left(\frac{\partial\mathcal{L}}{\partial x^\nu}\right)_\psi.$$

If the Lagrangian is explicitly independent of the space-time coordinates, i.e., $(\partial_\nu\mathcal{L})_\psi = 0$, the energy-momentum conservation law $\partial_\mu T^\mu{}_\nu = 0$ is written in terms of the energy-momentum tensor

$$T^\mu{}_\nu \equiv \mathcal{L}\,g^\mu{}_\nu - \frac{\partial\mathcal{L}}{\partial(\partial_\mu\psi)}\,\partial_\nu\psi. \qquad (9.14)$$

We note that the derivation of the energy-momentum conservation law is the same for classical and quantum fields. A similar procedure would lead to the conservation of angular momentum but this derivation is beyond the scope of the present Notes and we move on instead to an important conservation law in wave dynamics.

9.2.2.2   *Wave-Action Conservation Law*

Waves are known to exist on a great variety of media. When waves are supported by a spatially nonuniform or time-dependent medium, the conservation law of energy or momentum no longer apply and instead energy or momentum is transfered between the medium and the waves. There is however one conservation law which still applies and the quantity being conserved is known as *wave action*.

The derivation of a wave-action conservation law differs for classical fields and quantum fields. The difference is related to the fact that, whereas classical fields are generally represented by real-valued wave functions (i.e., $\psi^* = \psi$), the wave functions of quantum field theories are complex-valued (i.e., $\psi^* \neq \psi$).

The first step in deriving a wave-action conservation law in classical field theory involves transforming the real-valued wave function ψ into a complex-valued wave function ψ. Next, variations of ψ and its complex conjugate ψ^* are of the form

$$\delta\psi \equiv i\epsilon\,\psi \quad \text{and} \quad \delta\psi^* \equiv -i\epsilon\,\psi^*, \tag{9.15}$$

obtained by introducing an infinitesimal phase shift ϵ in ψ and ψ^*. Lastly, we transform the classical Lagrangian density $\mathcal{L}$ into a real-valued Lagrangian density $\mathcal{L}_R(\psi, \psi^*)$ such that $\delta\mathcal{L}_R \equiv 0$ (i.e., $\mathcal{L}_R$ is explicitly independent of the phase of the wave function ψ). The wave-action conservation law is, therefore, expressed in the form $\partial_\mu\,\mathcal{J}^\mu = 0$, where the wave-action four-density is

$$\mathcal{J}^\mu \equiv 2\,\mathrm{Im}\left[\psi\,\frac{\partial\mathcal{L}_R}{\partial(\partial_\mu\psi)}\right], \tag{9.16}$$

where $\mathrm{Im}[\cdots]$ denotes the imaginary part [i.e., $\mathrm{Im}(a^*b) = (a^*b - ab^*)/2i$].

The standard method in deriving the wave-action four-density (9.16) makes use of the *eikonal* representation for the real-valued wave field $\psi(\mathbf{x}, t)$:

$$\psi(\mathbf{x}, t) = \widetilde{\psi}(\epsilon\mathbf{x}, \epsilon t)\,e^{i\,\Theta(\epsilon\mathbf{x},\epsilon t)/\epsilon} + \widetilde{\psi}^*(\epsilon\mathbf{x}, \epsilon t)\,e^{-i\,\Theta(\epsilon\mathbf{x},\epsilon t)/\epsilon}, \tag{9.17}$$

where $\widetilde{\psi}$ denotes the complex-valued eikonal amplitude and Θ denotes the eikonal phase. The small parameter $\epsilon \ll 1$ indicates that the space-time gradient

$$\partial_\mu\psi = \left(i\,k_\mu\,\widetilde{\psi} + \epsilon\,\widetilde{\psi}_{,\mu}\right)e^{i\Theta/\epsilon} + \left(-i\,k_\mu\,\widetilde{\psi}^* + \epsilon\,\widetilde{\psi}^*_{,\mu}\right)e^{-i\Theta/\epsilon}$$

is expressed (to lowest order in ϵ) in terms of the wave four-vector

$$k_\mu = \epsilon^{-1}\,\partial_\mu\Theta \equiv \Theta_{,\mu} = (-\omega/c,\,\mathbf{k}), \tag{9.18}$$

where we used the eikonal relations (3.15). Hence, to lowest order in ϵ, the Lagrangian density $\mathcal{L}(\psi, \partial_\mu \psi)$ for a real-valued wave field ψ becomes the real-valued Lagrangian density $\mathcal{L}_R(\widetilde{\psi}; k_\mu)$ for the complex-valued wave field $\widetilde{\psi}$. The wave-action density (9.16) now simply becomes

$$\mathcal{J}^\mu \equiv \epsilon \, \frac{\partial \mathcal{L}_R}{\partial(\partial_\mu \Theta)} = \frac{\partial \mathcal{L}_R}{\partial k_\mu}, \tag{9.19}$$

and the wave-action conservation law becomes the Euler-Lagrange equation

$$\frac{\partial}{\partial x^\mu} \left(\frac{\partial \mathcal{L}_R}{\partial(\partial_\mu \Theta)} \right) = \frac{\partial \mathcal{L}_R}{\partial \Theta} \equiv 0,$$

which follows from the fact that $\mathcal{L}_R$ is independent of the eikonal phase Θ but not its space-time derivatives $\partial_\mu \Theta$.

9.3 Schroedinger's Equation*

A simple yet important example for a quantum field theory is provided by the Schroedinger equation for a spinless particle of mass m subjected to a real-valued potential energy function $U(\mathbf{x}, t)$. The Lagrangian density for the Schroedinger equation is given as

$$\mathcal{L}_R = -\frac{\hbar^2}{2m} |\nabla \psi|^2 + \frac{i\hbar}{2} \left(\psi^* \frac{\partial \psi}{\partial t} - \psi \frac{\partial \psi^*}{\partial t} \right) - U |\psi|^2. \tag{9.20}$$

The Schroedinger equation for ψ is derived as an Euler-Lagrange equation (9.10) in terms of ψ^*, where

$$\frac{\partial \mathcal{L}_R}{\partial(\partial_t \psi^*)} = -\frac{i\hbar}{2} \psi \;\; \rightarrow \;\; \frac{\partial}{\partial t} \left(\frac{\partial \mathcal{L}_R}{\partial(\partial_t \psi^*)} \right) = -\frac{i\hbar}{2} \frac{\partial \psi}{\partial t},$$

$$\frac{\partial \mathcal{L}_R}{\partial(\nabla \psi^*)} = -\frac{\hbar^2}{2m} \nabla \psi \;\; \rightarrow \;\; \nabla \cdot \left(\frac{\partial \mathcal{L}_R}{\partial(\nabla \psi^*)} \right) = -\frac{\hbar^2}{2m} \nabla^2 \psi,$$

$$\frac{\partial \mathcal{L}_R}{\partial \psi^*} = \frac{i\hbar}{2} \frac{\partial \psi}{\partial t} - U \psi.$$

By combining these derivatives, the Euler-Lagrange equation (9.10) for the Schroedinger Lagrangian (9.20) becomes

$$i\hbar \frac{\partial \psi}{\partial t} = -\frac{\hbar^2}{2m} \nabla^2 \psi + U \psi, \tag{9.21}$$

while the Schroedinger equation for ψ^* is as an Euler-Lagrange equation (9.10) in terms of ψ:

$$-i\hbar \frac{\partial \psi^*}{\partial t} = -\frac{\hbar^2}{2m} \nabla^2 \psi^* + U \psi^*, \tag{9.22}$$

which is simply the complex-conjugate equation of Eq. (9.21).

The energy-momentum conservation law for the Schroedinger equation (9.21) is now derived by Noether method. Because the potential $U(\mathbf{x}, t)$ is in general spatially nonuniform and time dependent, the energy-momentum contained in the wave function is not conserved and energy-momentum is exchanged between the wave function and the potential U. For example, the energy *transfer* equation is

$$\frac{\partial \mathcal{E}}{\partial t} + \nabla \cdot \mathbf{S} = |\psi|^2 \frac{\partial U}{\partial t}, \tag{9.23}$$

where the energy density $\mathcal{E}$ and energy density flux $\mathbf{S}$ are given explicitly as

$$\mathcal{E} = -\mathcal{L}_R + \frac{i\hbar}{2} \left(\psi^* \frac{\partial \psi}{\partial t} - \psi \frac{\partial \psi^*}{\partial t} \right)$$

$$\mathbf{S} = -\frac{\hbar^2}{2m} \left(\frac{\partial \psi}{\partial t} \nabla \psi^* + \frac{\partial \psi^*}{\partial t} \nabla \psi \right).$$

The momentum transfer equation, on the other hand, is

$$\frac{\partial \mathbf{P}}{\partial t} + \nabla \cdot \mathsf{T} = -|\psi|^2 \nabla U, \tag{9.24}$$

where the momentum density $\mathbf{P}$ and momentum density tensor T are given explicitly as

$$\mathbf{P} = \frac{i\hbar}{2} \left(\psi \nabla \psi^* - \psi^* \nabla \psi \right)$$

$$\mathsf{T} = \mathcal{L}_R \, \mathsf{I} + \frac{\hbar^2}{2m} \left(\nabla \psi^* \nabla \psi + \nabla \psi \nabla \psi^* \right).$$

Note that Eqs. (9.23) and (9.24) are both exact equations for any time-dependent, nonuniform potential $U(\mathbf{x}, t)$.

Whereas energy-momentum is transfered between the wave function ψ and the potential V, the amount of wave-action contained in the wave function is conserved. Indeed, the wave-action conservation law is

$$\frac{\partial \mathcal{J}}{\partial t} + \nabla \cdot \mathbf{J} = 0, \tag{9.25}$$

where, according to Eq. (9.16), the wave-action density $\mathcal{J}$ and wave-action density flux $\mathbf{J}$ are

$$\mathcal{J} = \hbar |\psi|^2 \quad \text{and} \quad \mathbf{J} = \frac{\hbar^2}{m} \operatorname{Im} (\psi^* \nabla \psi). \tag{9.26}$$

Thus wave-action conservation law is none other than the law of conservation of probability associated with the normalization condition

$$\int |\psi|^2 \, d^3x = 1$$

for bounds states or the conservation of the number of quanta in a scattering problem.

Lastly, by substituting the *ansatz*

$$\psi \equiv \sqrt{\rho} \, \exp(iS/\hbar) \tag{9.27}$$

in the Schroedinger Lagrangian density (9.20), where $\rho > 0$ and S are real-valued functions, we can easily obtain the Lagrangian density

$$\mathcal{L}_C = -\rho \left(\frac{\partial S}{\partial t} + \frac{|\nabla S|^2}{2m} + U \right)$$

in the classical limit $\hbar \to 0$. The variational principle

$$\delta \int \mathcal{L}_C(\rho; S, \partial_t S, \nabla S) \, dt = 0$$

with respect to variations $\delta\rho$ then yields the Hamilton-Jacobi equation (3.8)

$$\frac{\partial S}{\partial t} + \frac{|\nabla S|^2}{2m} + U \equiv \frac{\partial S}{\partial t} + H(\mathbf{x}, \nabla S; t) = 0,$$

where H is the Hamiltonian function with momentum $\mathbf{p} \equiv \nabla S$ defined in terms of S (see Table 3.1). We thus see the explicit connection between the Hamilton-Jacobi equation for classical particle dynamics and the Schroedinger equation (9.21) for quantum mechanics. The Euler-Lagrange equation for ρ (corresponding to variations δS), on the other hand, yields the conservation law

$$\frac{\partial \rho}{\partial t} + \nabla \cdot \left(\rho \, \frac{\nabla S}{m} \right) = 0,$$

which is identical to the wave-action conservation law (9.25), with $\mathcal{J} \equiv \hbar \rho$ and $\mathbf{J} \equiv \mathcal{J} \nabla S / m$.

9.4 Euler Equations for a Perfect Fluid

Given Euler's role in the development of the Calculus of Variations (in Chap. 1), it is extremely fitting to end this textbook on Lagrangian Mechanics by considering the Euler equations of motion for a perfect fluid

$$\frac{\partial \rho}{\partial t} + \nabla \cdot (\rho \mathbf{u}) = 0, \tag{9.28}$$

$$\rho \left(\frac{\partial}{\partial t} + \mathbf{u} \cdot \nabla \right) \mathbf{u} = -\nabla p, \tag{9.29}$$

$$\frac{\partial S}{\partial t} + \mathbf{u} \cdot \nabla S = 0. \tag{9.30}$$

Equation (9.28) represents the particle conservation law, where $\rho(\mathbf{x}, t)$ denotes the mass density of the fluid and $\mathbf{u}(\mathbf{x}, t)$ denotes the fluid velocity. Equation (9.29) represents Newton's Second Law for the perfect fluid, which states that the fluid moves under the influence of a pressure gradient force $-\nabla p$. Equation (9.30) represents the conservation of entropy, which states that entropy is *advected* with the fluid.

According to the First Law of Thermodynamics, the mass density ρ and the entropy S (per unit mass) of the fluid can be used as independent variables so that a change $\delta\varepsilon(\rho, S)$ in the internal energy (per unit mass) of the fluid can be expressed as

$$\delta\varepsilon = T\,\delta S - p\,\delta\rho^{-1}, \tag{9.31}$$

where T and p denote the temperature and pressure of the fluid.

9.4.1 *Lagrangian Formulation*

The Euler-Lagrange formulation of the Euler fluid equations (9.28)-(9.30) is based on the Lagrangian density

$$\mathcal{L} = \frac{1}{2}\rho\,|\mathbf{u}|^2 - \rho\varepsilon + \phi\left(\frac{\partial\rho}{\partial t} + \nabla\cdot\rho\mathbf{u}\right) - \rho\lambda\,\frac{dS}{dt}, \tag{9.32}$$

where ϕ and λ are Lagrange multipliers used to enforce the particle conservation law (9.28) and the constraint (9.30) that entropy is advected by the fluid. Here, the variational fields are represented by the seven-component field $\psi^a = (\rho, \mathbf{u}, S; \phi, \lambda)$ and the Euler-Lagrange equation for each component ψ^a is

$$\frac{\partial}{\partial t}\left(\frac{\partial\mathcal{L}}{\partial(\partial_t\psi^a)}\right) + \nabla\cdot\left(\frac{\partial\mathcal{L}}{\partial(\nabla\psi^a)}\right) = \frac{\partial\mathcal{L}}{\partial\psi^a}. \tag{9.33}$$

We note that the Lagrangian density (9.32) does not have space-time derivatives for all fields. For example, it is immediately obvious that we recover the constraint equations (9.28) and (9.30) from variations of the Lagrangian density with respect to the Lagrange multipliers ϕ and λ (i.e., $\partial\mathcal{L}/\partial\phi = 0 = \delta\mathcal{L}/\partial\lambda$).

The Euler-Lagrange equation for the mass density ρ is

$$\frac{d\phi}{dt} \equiv \frac{\partial\phi}{\partial t} + \mathbf{u}\cdot\nabla\phi = \frac{1}{2}\,|\mathbf{u}|^2 - h, \tag{9.34}$$

where $h \equiv \partial(\rho\varepsilon)/\partial\rho = \varepsilon + p/\rho$ is the *enthalpy* of the fluid. The Euler-Lagrange equation for the entropy S is

$$\frac{d\lambda}{dt} = T. \tag{9.35}$$

Lastly, the Euler-Lagrange equation for the fluid velocity **u** yields

$$\mathbf{u} = \nabla\phi + \lambda\,\nabla S, \tag{9.36}$$

which introduces a decomposition of the fluid velocity in terms of a curl-free term (i.e., $\nabla \times \nabla\phi = 0$) and a term that is proportional to the entropy gradient.

We now show that the equation of motion (9.29) is contained in Eqs. (9.34)-(9.36) as follows. First, we write the partial time derivative of Eq. (9.36)

$$\frac{\partial\mathbf{u}}{\partial t} = \nabla\frac{\partial\phi}{\partial t} + \frac{\partial\lambda}{\partial t}\,\nabla S + \lambda\,\nabla\frac{\partial S}{\partial t}$$

$$= \nabla\left(\frac{1}{2}\,|\mathbf{u}|^2 - h - \mathbf{u}\cdot\nabla\phi\right) + (T - \mathbf{u}\cdot\nabla\lambda)\,\nabla S - \lambda\,\nabla(\mathbf{u}\cdot\nabla S).$$

By rearranging terms, we find

$$\frac{\partial\mathbf{u}}{\partial t} = \nabla\mathbf{u}\cdot(\mathbf{u} - \nabla\phi - \lambda\,\nabla S) - \mathbf{u}\cdot\nabla(\nabla\phi + \lambda\,\nabla S)$$
$$- \nabla h + T\,\nabla S.$$

Lastly, by using Eq. (9.36) and the identity $T\,\nabla S - \nabla h \equiv -\rho^{-1}\nabla p$, we recover Eq. (9.29).

9.4.2   *Energy-Momentum Conservation Laws*

We now derive the energy-momentum conservation laws for the Euler fluid equations (9.28)-(9.30). The Noether equation for the Lagrangian density (9.32) is expressed as

$$\delta\mathcal{L} = \frac{\partial}{\partial t}\left(\phi\,\delta\rho - \rho\lambda\,\delta S\right) + \nabla\cdot\left[(\delta\rho\,\mathbf{u} + \rho\,\delta\mathbf{u})\,\phi - \rho\mathbf{u}\,\lambda\,\delta S\right]. \tag{9.37}$$

Here, the variations $(\delta\rho, \delta\mathbf{u}, \delta S)$ are expressed in terms of space-time translations $\boldsymbol{\xi} \equiv \delta\mathbf{x} - \mathbf{u}\,\delta t$ as

$$\delta\rho = -\nabla\cdot(\boldsymbol{\xi}\,\rho), \tag{9.38}$$

$$\delta\mathbf{u} = \frac{\partial\boldsymbol{\xi}}{\partial t} + \mathbf{u}\cdot\nabla\boldsymbol{\xi} - \boldsymbol{\xi}\cdot\nabla\mathbf{u}, \tag{9.39}$$

$$\delta S = -\boldsymbol{\xi}\cdot\nabla S, \tag{9.40}$$

which satisfy the constraint equations

$$\frac{\partial\delta\rho}{\partial t} = -\nabla\cdot(\delta\rho\,\mathbf{u} + \rho\,\delta\mathbf{u}), \tag{9.41}$$

$$\frac{\partial\delta S}{\partial t} = -\delta\mathbf{u}\cdot\nabla S - \mathbf{u}\cdot\nabla\delta S. \tag{9.42}$$

By substituting Eqs. (9.38)-(9.40) on the right side of the Noether equation (9.37), we obtain

$$\delta\mathcal{L} = \frac{\partial}{\partial t}\left(\rho\,\boldsymbol{\xi}\cdot\mathbf{u}\right) + \nabla\cdot\left[\mathbf{u}\left(\rho\,\boldsymbol{\xi}\cdot\mathbf{u}\right) - \rho\boldsymbol{\xi}\left(\frac{1}{2}\left|\mathbf{u}\right|^2 - h\right)\right], \quad (9.43)$$

after carrying out several cancellations as well as using the identity

$$\nabla\cdot\left[\rho\phi\left(\mathbf{u}\,\boldsymbol{\xi} - \boldsymbol{\xi}\,\mathbf{u}\right)\right] = \nabla\times\left(\rho\phi\,\boldsymbol{\xi}\times\mathbf{u}\right),$$

and using Eq. (9.34) for $d\phi/dt$.

First, the energy conservation law

$$\frac{\partial\mathcal{E}}{\partial t} + \nabla\cdot\mathbf{S} = 0 \quad (9.44)$$

is associated with time-translation symmetry for which $\delta\mathcal{L} = -\delta t\,\partial\mathcal{L}/\partial t$, where the energy density and energy-density flux are

$$\mathcal{E} = \rho\left|\mathbf{u}\right|^2 - \mathcal{L} = \rho\left(\frac{1}{2}\left|\mathbf{u}\right|^2 + \varepsilon\right), \quad (9.45)$$

$$\mathbf{S} = \rho\mathbf{u}\left(\frac{1}{2}\left|\mathbf{u}\right|^2 + \varepsilon\right) + p\,\mathbf{u}. \quad (9.46)$$

Second, the momentum conservation

$$\frac{\partial\mathbf{P}}{\partial t} + \nabla\cdot\mathsf{T} = 0 \quad (9.47)$$

is associated with space-translation symmetry for which $\delta\mathcal{L} = -\delta\mathbf{x}\cdot\nabla\mathcal{L}$, where the momentum density and stress tensor are

$$\mathbf{P} = \rho\mathbf{u} \quad\text{and}\quad \mathsf{T} = \rho\,\mathbf{u}\,\mathbf{u} + p\,\mathsf{I}. \quad (9.48)$$

The Euler fluid equations also possess a wave-action conservation law, which requires us to introduce a fluid reference state on which waves propagate.

We note in closing that the Euler fluid equations (9.28)-(9.30) possess a different Lagrangian formulation (see problem 5) that makes use of constrained variations for the fluid fields $(\rho, \mathbf{u}, S)$ without the use of Lagrange multipliers (ϕ, λ).

9.5 Problems

1. When we insert the ansatz (9.27) into the Schroedinger equation (9.21), we obtain two equations defined as the real and imaginary parts of the resulting Schroedinger equation. Derive these two equations and give their interpretations in the classical limit $\hbar \to 0$.

2. Show that Eqs. (9.34)-(9.36) have the Euler-Lagrange form (9.33).

3. Show that Eqs. (9.38)-(9.40) satisfy the constraint equations (9.41)-(9.42).

4. Consider a perfect fluid under the influence of an external force (e.g., gravity) and subject to the equation of motion

$$\rho \left(\frac{\partial}{\partial t} + \mathbf{u} \cdot \nabla \right) \mathbf{u} = -\nabla p - \rho \nabla \Phi,$$

where $\Phi(\mathbf{x}, t)$ denotes the scalar potential (per unit mass) associated with the external force. Show that the new equation of motion can be derived from the new Lagrangian density

$$\mathcal{L}' = \frac{1}{2} \rho |\mathbf{u}|^2 - \rho (\varepsilon + \Phi) + \phi \left(\frac{\partial \rho}{\partial t} + \nabla \cdot \rho \mathbf{u} \right) - \rho \lambda \frac{dS}{dt}.$$

5. * Show that the Euler fluid equations (9.28)-(9.30) can be formulated in terms of a constrained variational principle, with the Lagrangian density

$$\mathcal{L} = \frac{1}{2} \rho |\mathbf{u}|^2 - \rho \, \varepsilon(\rho, S),$$

where the constrained variations

$$\delta \rho = -\nabla \cdot (\rho \boldsymbol{\xi}),$$

$$\delta \mathbf{u} = \frac{\partial \boldsymbol{\xi}}{\partial t} + \mathbf{u} \cdot \nabla \boldsymbol{\xi} - \boldsymbol{\xi} \cdot \nabla \mathbf{u},$$

$$\delta S = -\boldsymbol{\xi} \cdot \nabla S,$$

are expressed in terms of the virtual fluid displacement $\boldsymbol{\xi}$.

Appendix A

Basic Mathematical Methods

Appendix A introduces, first, an explicit derivation of the Frenet-Serret formulas for an arbitrary curve in three-dimensional space used in Chapters 1 and 2. Next, some basic concepts in linear algebra that a student may have acquired before taking this course are summarized. Hopefully, this material will assist the student in following the presentation in Chapters 7 and 8. Lastly, some general comments are made concerning the numerical analysis of the nonlinear (and coupled) differential equations presented in this textbook.

A.1 Frenet-Serret Formulas

A.1.1 *General Formulas*

Consider a curve

$$\mathbf{r}(t) = x(t)\widehat{\mathbf{x}} + y(t)\widehat{\mathbf{y}} + z(t)\widehat{\mathbf{z}} \tag{A.1}$$

in three-dimensional space parameterized by time t. The infinitesimal length element along the curve $ds(t) = v(t)\, dt$ is also parameterized by time t, with $v(t) \equiv |\dot{\mathbf{r}}|$ denoting the speed along the curve.

The Frenet-Serret formulas associated with the curvature κ and torsion τ of the curve (A.1) are defined in terms of the right-handed set of unit vectors $(\widehat{\mathbf{t}}, \widehat{\mathbf{n}}, \widehat{\mathbf{b}})$, where $\widehat{\mathbf{t}}$ denotes the *tangent* unit vector, $\widehat{\mathbf{n}}$ denotes the *normal* unit vector, and $\widehat{\mathbf{b}}$ denotes the *binormal* unit vector. First, by definition, the tangent unit vector is

$$\widehat{\mathbf{t}} \equiv \frac{d\mathbf{r}}{ds} = \frac{\dot{\mathbf{r}}(t)}{v(t)}. \tag{A.2}$$

The definitions of the curvature κ and the normal unit vector $\widehat{\mathbf{n}}$ are defined as

$$\frac{d\widehat{t}}{ds} = \frac{\ddot{r}}{v^2} - \frac{\dot{v}}{v^2}\,\widehat{t} = \frac{\widehat{t}}{v^2} \times (\ddot{r} \times \widehat{t})$$

$$= \left(\frac{\dot{r} \times \ddot{r}}{v^3}\right) \times \widehat{t} = \kappa\,\left(\widehat{b} \times \widehat{t}\right) \equiv \kappa\,\widehat{n}, \tag{A.3}$$

where $\dot{v} \equiv d|\dot{r}|/dt = \widehat{t}\cdot\ddot{r}$, so that the curvature is defined as

$$\kappa \equiv \frac{|\dot{r} \times \ddot{r}|}{v^3}, \tag{A.4}$$

while the normal and binormal unit vectors are defined as

$$\widehat{n} \equiv \frac{\dot{r} \times (\ddot{r} \times \dot{r})}{\kappa\,v^4} = \widehat{t} \times \left(\frac{\ddot{r} \times \dot{r}}{|\ddot{r} \times \dot{r}|}\right), \tag{A.5}$$

and

$$\widehat{b} \equiv \frac{\dot{r} \times \ddot{r}}{|\dot{r} \times \ddot{r}|}. \tag{A.6}$$

Hence, the curve (A.1) exhibits curvature if its velocity $\dot{r}$ and acceleration $\ddot{r}$ are not colinear.

Next, we obtain the following expression for the derivative of the normal unit vector (A.5):

$$\frac{d\widehat{n}}{ds} = \frac{d\widehat{t}}{ds} \times \left(\frac{\ddot{r} \times \dot{r}}{|\ddot{r} \times \dot{r}|}\right) + \widehat{t} \times \frac{d}{ds}\left(\frac{\ddot{r} \times \dot{r}}{|\ddot{r} \times \dot{r}|}\right)$$

$$= \left[\widehat{t} \times \left(\frac{\ddot{r} \times \dot{r}}{v^3}\right)\right] \times \left(\frac{\ddot{r} \times \dot{r}}{|\ddot{r} \times \dot{r}|}\right) + \frac{\widehat{t}}{v} \times \left[\widehat{b} \times \left(\frac{\left(\dddot{r} \times \dot{r}\right) \times \widehat{b}}{|\ddot{r} \times \dot{r}|}\right)\right]$$

$$= \kappa\,\left(\widehat{t} \times \widehat{b}\right) \times \widehat{b} + \left(\frac{\widehat{t}\cdot\left[\left(\dddot{r} \times \dot{r}\right) \times \widehat{b}\right]}{v\,|\ddot{r} \times \dot{r}|}\right)\widehat{b} \equiv -\,\kappa\widehat{t} + \tau\widehat{b}, \tag{A.7}$$

where the torsion

$$\tau \equiv \frac{\widehat{t}\cdot\left[\left(\dddot{r} \times \dot{r}\right) \times \widehat{b}\right]}{v\,|\ddot{r} \times \dot{r}|} = \frac{\widehat{n}\cdot\left(\dddot{r} \times \dot{r}\right)}{\kappa\,v^4} = \frac{\dot{r}\cdot\left(\ddot{r} \times \dddot{r}\right)}{\kappa^2\,v^6} \tag{A.8}$$

is defined in terms of the triple product $\dot{r}\cdot(\ddot{r} \times \dddot{r})$. Hence, the torsion requires that the rate of change of acceleration $\dddot{r} = d\ddot{r}/dt$ (known as *jerk*) along the curve have a nonvanishing component perpendicular to the plane constructed by the velocity $\dot{r}$ and the acceleration $\ddot{r}$.

Lastly, we obtain the expression for the derivative of the binormal unit vector (A.6):

$$\frac{d\widehat{b}}{ds} = \frac{\widehat{b} \times \left[\left(\dot{r} \times \dddot{r}\right) \times \widehat{b}\right]}{v\,|\dot{r} \times \ddot{r}|} = \alpha\widehat{t} + \beta\widehat{n} \equiv -\,\tau\,\widehat{n}, \tag{A.9}$$

where

$$\alpha \equiv \widehat{t} \cdot \frac{d\widehat{b}}{ds} = \frac{\widehat{t} \cdot \left(\dot{r} \times \dddot{r}\right)}{\kappa \, v^4} \equiv 0,$$

and

$$\beta \equiv \widehat{n} \cdot \frac{d\widehat{b}}{ds} = \frac{\widehat{n} \cdot \left(\dot{r} \times \dddot{r}\right)}{\kappa \, v^4} = \left[\frac{\dot{r} \times \left(\ddot{r} \times \dot{r}\right)}{\kappa^2 \, v^8} \right] \cdot \left(\dot{r} \times \dddot{r}\right)$$

$$= \frac{\ddot{r} \cdot \left(\dot{r} \times \dddot{r}\right)}{\kappa^2 \, v^6} = -\,\tau.$$

The equations (A.3), (A.7), and (A.9) are refered to as the Frenet-Serret formulas, which describes the evolution of the unit vectors $(\widehat{t}, \widehat{n}, \widehat{b})$ along the curve (A.1) in terms of the curvature (A.4) and the torsion (A.8). Note that by introducing the *Darboux* vector (Gaston Darboux, 1842-1917)

$$\boldsymbol{\omega} \equiv \tau\widehat{t} + \kappa\widehat{b}, \tag{A.10}$$

the Frenet-Serret equations (A.3), (A.7), and (A.9) may be written as

$$\frac{d\widehat{e}_i}{ds} \equiv \boldsymbol{\omega} \times \widehat{e}_i,$$

where $\widehat{e}_i = (\widehat{t}, \widehat{n}, \widehat{b})$ denotes a component of the so-called Frenet *frame*. Hence, curvature is a measure of the rotation of the Frenet frame about the binormal unit vector $\widehat{b}$, while torsion is the measure of the rotation of the Frenet frame about the tangent unit vector $\widehat{t}$.

A.1.2 *Frenet-Serret Formulas for Helical Path*

As a simple application of the Frenet-Serret formulas, we apply them to the helical path

$$r(\theta) = a\,(\cos\theta\,\widehat{x} + \sin\theta\,\widehat{y}) + b\theta\,\widehat{z} \tag{A.11}$$

parameterized by the angle θ. Here, the distance s along the helix is simply given as $s = c\,\theta$, where $c = \sqrt{a^2 + b^2}$. Hence, the tangent unit vector is defined as

$$\widehat{t} = \frac{dr}{ds} = \cos\alpha\,(-\sin\theta\,\widehat{x} + \cos\theta\,\widehat{y}) + \sin\alpha\,\widehat{z}, \tag{A.12}$$

where $(a, b) \equiv (c\cos\alpha, c\sin\alpha)$ and $0 \le \alpha < \pi/2$ denotes the pitch of the helix (e.g., a circle is a helical path with pitch $\alpha = 0$). Next, the derivative of the tangent unit vector yields

$$\frac{d\widehat{t}}{ds} = -\frac{\cos\alpha}{c}\,(\cos\theta\,\widehat{x} + \sin\theta\,\widehat{y}) \equiv \kappa\,\widehat{n}, \tag{A.13}$$

so that the normal unit vector is

$$\widehat{n} = - (\cos\theta\,\widehat{x} + \sin\theta\,\widehat{y}) \tag{A.14}$$

and the curvature is $\kappa = c^{-1}\cos\alpha$ (i.e., a circle of radius a, with pitch $\alpha = 0$, has a scalar curvature $\kappa = a^{-1}$). Lastly, the binormal vector is

$$\widehat{b} = \widehat{t}\times\widehat{n} = \sin\alpha\,(-\sin\theta\,\widehat{x} + \cos\theta\,\widehat{y}) + \cos\alpha\,\widehat{z}, \tag{A.15}$$

so that its derivative yields

$$\frac{d\widehat{b}}{ds} = \frac{\sin\alpha}{c}\,(\cos\theta\,\widehat{x} + \sin\theta\,\widehat{y}) \equiv -\tau\,\widehat{n}, \tag{A.16}$$

where the torsion is $\tau = c^{-1}\sin\alpha$. We can now easily verify that

$$\frac{d\widehat{n}}{ds} = \frac{1}{c}\,(\sin\theta\,\widehat{x} - \cos\theta\,\widehat{y}) \equiv -\kappa\,\widehat{t} + \tau\,\widehat{b}.$$

We point out that for a two-dimensional curve $\mathbf{r} = x(s)\,\widehat{x} + y(s)\,\widehat{y}$, we find

$$\widehat{t} = \frac{d\mathbf{r}}{ds} = x'\,\widehat{x} + y'\,\widehat{y} \equiv \cos\phi\,\widehat{x} + \sin\phi\,\widehat{y},$$

where $\phi(s)$ denotes the tangential angle. With this definition, we readily show that the curvature is defined as

$$\kappa \equiv \left|\frac{d\widehat{t}}{ds}\right| = \frac{d\phi}{ds}.$$

Exercise: Calculate the Darboux vector (A.10) for the helical path (A.11).

A.2   Linear Algebra

A fundamental object in linear algebra is the $m \times n$ matrix A with components (labeled A_{ij}) distributed on m rows ($i = 1, 2, ..., m$) and n columns ($j = 1, 2, ..., n$); for simplicity of notation, we write $\mathsf{A}_{(m\times n)}$ when we want to specify the *order* of the matrix A and we say that a matrix is *square* if $m = n$.

A.2.1   *Matrix Algebra*

We begin with a discussion of general properties of matrices and later focus our attention on square matrices (in particular 2×2 matrices). First, we can add (or subtract) two matrices only if they are of the same order; hence, the matrix $\mathsf{C} = \mathsf{A} \pm \mathsf{B}$ has components $C_{ij} = A_{ij} \pm B_{ij}$. Next, we can multiply a matrix A by a scalar a and obtain the new matrix $B = a\,A$ with components $B_{ij} = a\,A_{ij}$. Lastly, we introduce the *transpose* operation (denoted $^\top$): $A \to A^\top$ such that $(A^\top)_{ij} \equiv A_{ji}$, i.e., the transpose of a $m \times n$ matrix is a $n \times m$ matrix. Note that the vector

$$\mathbf{v} = \begin{pmatrix} v_1 \\ \vdots \\ v_n \end{pmatrix}$$

is a $n \times 1$ matrix while its transpose $\mathbf{v}^\top = (v_1, ..., v_n)$ is a matrix of order $1 \times n$. With this definition, we now introduce the operation of matrix multiplication

$$C_{(m \times k)} = A_{(m \times n)} \cdot B_{(n \times k)},$$

where $C_{(m \times k)}$ is a new matrix of order $m \times k$ with components

$$C_{ij} = \sum_{\ell=1}^{n} A_{i\ell}\, B_{\ell j}.$$

Note that the matrix multiplication

$$\mathbf{u}^\top \cdot \mathbf{v} = \sum_{i=1}^{n} u_i\, v_i = \mathbf{u} \cdot \mathbf{v}$$

coincides with the standard dot product of two vectors.

The remainder of this Section will now exclusively deal with square matrices. First, we introduce two important operations on square matrices: the determinant $\det(A)$ and the trace $\mathrm{Tr}(A)$ defined, respectively, as

$$\det(A) \equiv \sum_{i=1}^{n} (-1)^{i+j}\, A_{ij}\, \mathrm{ad}_{ij} \equiv \sum_{j=1}^{n} (-1)^{i+j}\, A_{ij}\, \mathrm{ad}_{ij}, \qquad (A.17)$$

$$\mathrm{Tr}(A) \equiv \sum_{i=1}^{n} A_{ii}, \qquad (A.18)$$

where ad_{ij} denotes the determinant of the *reduced* matrix obtained by removing the ith-row and j^{th}-column from A and the index j is fixed in the first expression in Eq. (A.17), while the index i is fixed in the second expression. Next, we say that the matrix A is invertible if its determinant $\Delta \equiv \det(A)$ does not vanish and we define the inverse A^{-1} with components

$$(A^{-1})_{ij} \equiv \frac{(-1)^{i+j}}{\Delta}\, \mathrm{ad}_{ji},$$

which, thus, satisfies the identity relation

$$\mathsf{A} \cdot \mathsf{A}^{-1} = \mathbf{I} = \mathsf{A}^{-1} \cdot \mathsf{A},$$

where $\mathbf{I}$ denotes the $n \times n$ identity matrix. Note, here, that the matrix multiplication

$$\mathsf{A} \cdot \mathsf{B} \neq \mathsf{B} \cdot \mathsf{A}$$

of two matrices A and B is generally not commutative (for $\mathsf{A}, \mathsf{B} \neq \mathbf{I}$).

Lastly, fundamental properties of a square $n \times n$ matrix A are discussed in terms of its eigenvalues $(\lambda_1, ..., \lambda_n)$ and eigenvectors $(\mathbf{e}_1, ..., \mathbf{e}_n)$ which satisfy the eigenvalue equation

$$\mathsf{A} \cdot \mathbf{e}_i = \lambda_i \, \mathbf{e}_i, \tag{A.19}$$

for $i = 1, ..., n$. Here, the determinant and the trace of the $n \times n$ matrix A are expressed in terms of its eigenvalues $(\lambda_1, ..., \lambda_n)$ as

$$\det(\mathsf{A}) = \lambda_1 \times ... \times \lambda_n \quad \text{and} \quad \text{Tr}(\mathsf{A}) = \lambda_1 + ... + \lambda_n.$$

In order to continue our discussion of this important problem, we now focus our attention on 2×2 matrices.

A.2.2 *Eigenvalue Analysis of a 2 × 2 Matrix*

Consider the 2×2 matrix

$$\mathsf{M} = \begin{pmatrix} a & b \\ c & d \end{pmatrix}, \tag{A.20}$$

where (a, b, c, d) are arbitrary real (or complex) numbers and introduce the following two matrix *invariants*:

$$\Delta \equiv \det(\mathsf{M}) = a\,d - b\,c \quad \text{and} \quad \sigma \equiv \text{Tr}(\mathsf{M}) = a + d, \tag{A.21}$$

which denote the determinant and the trace of matrix M, respectively.

A.2.2.1 *Eigenvalues of* M

The *eigenvalues* λ and *eigenvectors* $\mathbf{e}$ of matrix M are defined by the eigenvalue equation

$$\mathsf{M} \cdot \mathbf{e} = \lambda \mathbf{e}. \tag{A.22}$$

This equation has nontrivial solutions only if the determinant of the matrix $\mathsf{M} - \lambda \mathbf{I}$ vanishes (where $\mathbf{I}$ denotes the 2×2 identity matrix). This vanishing determinant yields the characteristic quadratic polynomial:

$$\det(\mathsf{M} - \lambda \mathbf{I}) = (a - \lambda)(d - \lambda) - b\,c \equiv \lambda^2 - \sigma \lambda + \Delta = 0, \tag{A.23}$$

and the eigenvalues $\lambda_\pm$ are obtained as the roots of this characteristic polynomial:

$$\lambda_\pm = \frac{\sigma}{2} \pm \sqrt{\frac{\sigma^2}{4} - \Delta}. \tag{A.24}$$

Here, we note that the matrix invariants (σ, Δ) are related to the eigenvalues $\lambda_\pm$:

$$\lambda_+ + \lambda_- \equiv \sigma \quad \text{and} \quad \lambda_+ \cdot \lambda_- \equiv \Delta. \tag{A.25}$$

Lastly, the eigenvalues are said to be *degenerate* if $\lambda_+ = \lambda_- \equiv \sigma/2$, i.e.,

$$\Delta = \frac{\sigma^2}{4} \quad \text{or} \quad bc = -\left(\frac{a-d}{2}\right)^2.$$

A.2.2.2   *Eigenvectors of* M

Next, the eigenvectors $e_\pm$ associated with the eigenvalues $\lambda_\pm$ are constructed from the eigenvalue equations $M \cdot e_\pm = \lambda_\pm e_\pm$, which yield the general solutions

$$e_\pm \equiv \begin{pmatrix} 1 \\ \mu_\pm \end{pmatrix} \epsilon_\pm, \tag{A.26}$$

where $\epsilon_\pm$ denotes an arbitrary constant and

$$\mu_\pm = \frac{\lambda_\pm - a}{b} = \frac{c}{\lambda_\pm - d}. \tag{A.27}$$

The normalization of the eigenvectors $e_\pm$ ($|e_\pm| = 1$), for example, can be achieved by choosing

$$\epsilon_\pm = \frac{1}{\sqrt{1 + (\mu_\pm)^2}}.$$

We note that the eigenvectors $e_\pm$ are not automatically orthogonal to each other (i.e., the dot product $e_+ \cdot e_-$ may not vanish). Indeed, we find

$$e_+ \cdot e_- = \epsilon_+ \epsilon_- (1 + \mu_+ \mu_-), \tag{A.28}$$

where

$$1 + \mu_+ \mu_- = 1 + \frac{1}{b^2} (\lambda_+ - a)(\lambda_- - a) = 1 + \left(\frac{a-d}{2b}\right)^2 - \frac{1}{b^2}\left(\frac{\sigma^2}{4} - \Delta\right),$$

whose sign is indefinite. By using the Gram-Schmidt orthogonalization procedure, however, we may construct two orthogonal vectors (e_1, e_2):

$$\left. \begin{aligned} e_1 &= \alpha\, e_+ + \beta\, e_- \\ e_2 &= \gamma\, e_+ + \delta\, e_- \end{aligned} \right\}, \tag{A.29}$$

where the coefficients $(\alpha, \beta, \gamma, \delta)$ are chosen to satisfy the orthogonalization condition $\mathbf{e}_1 \cdot \mathbf{e}_2 \equiv 0$; it is important to note that the vectors $(\mathbf{e}_1, \mathbf{e}_2)$ are not themselves eigenvectors of the matrix M. For example, we may choose $\alpha = 1 = \delta$, $\beta = 0$, and

$$\gamma = -\left(\frac{\mathbf{e}_+ \cdot \mathbf{e}_-}{|\mathbf{e}_+|^2}\right),$$

which corresponds to choosing $\mathbf{e}_1 = \mathbf{e}_+$ and constructing $\mathbf{e}_2$ as the component of $\mathbf{e}_-$ that is orthogonal to $\mathbf{e}_+$.

Lastly, we point out that any two-dimensional vector $\mathbf{u}$ may be decomposed in terms of the eigenvectors $\mathbf{e}_\pm$:

$$\mathbf{u} = \sum_{i=\pm} u_i \, \mathbf{e}_i \equiv \sum_{i=\pm} \left(\frac{\mathbf{u} \cdot \mathbf{e}_i}{|\mathbf{e}_i|^2}\right) \mathbf{e}_i, \qquad (A.30)$$

where we assumed, here, that the eigenvectors are orthogonal to each other. Furthermore, the *transformation* $\mathsf{M} \cdot \mathbf{u}$ generates a new vector

$$\mathbf{v} = \mathsf{M} \cdot \mathbf{u} = \sum_{i=\pm} u_i \, \mathsf{M} \cdot \mathbf{e}_i \equiv \sum_{i=\pm} v_i \, \mathbf{e}_i,$$

where the components of $\mathbf{v}$ are $v_i \equiv u_i \, \lambda_i$.

A.2.2.3 *Inverse of matrix* M

The matrix (A.20) has an inverse, denoted M^{-1}, if its determinant Δ does not vanish. In this nonsingular case, we easily find

$$\mathsf{M}^{-1} = \frac{1}{\Delta}\begin{pmatrix} d & -b \\ -c & a \end{pmatrix}, \qquad (A.31)$$

so that $\mathsf{M}^{-1} \cdot \mathsf{M} = \mathsf{I} = \mathsf{M} \cdot \mathsf{M}^{-1}$. The determinant of M^{-1}, denoted Δ', is

$$\Delta' = \frac{d\,a - b\,c}{\Delta^2} \equiv \frac{1}{\Delta} = \frac{1}{\lambda_+ \cdot \lambda_-},$$

while its trace, denoted σ', is

$$\sigma' = \frac{d + a}{\Delta} \equiv \frac{\sigma}{\Delta} = \frac{1}{\lambda_+} + \frac{1}{\lambda_-}.$$

Hence, the eigenvalues of the inverse matrix (A.31) are

$$\lambda'_\pm \equiv \frac{1}{\lambda_\pm} = \frac{\lambda_\mp}{\Delta},$$

and its eigenvectors $\bar{\mathbf{e}}_\pm$ are identical to $\mathbf{e}_\pm$ since

$$\mathsf{M} \cdot \mathbf{e}_\pm = \lambda_\pm \, \mathbf{e}_\pm \;\; \to \;\; \mathbf{e}_\pm = \lambda_\pm \, \mathsf{M}^{-1} \cdot \mathbf{e}_\pm \;\; \to \;\; \mathsf{M}^{-1} \cdot \mathbf{e}_\pm = \lambda_\pm^{-1} \mathbf{e}_\pm \equiv \lambda'_\pm \, \mathbf{e}_\pm.$$

We note that once the inverse M^{-1} of a matrix M is known, then any inhomogeneous linear system of equations of the form $\mathsf{M} \cdot \mathbf{u} = \mathbf{v}$ may be solved as $\mathbf{u} \equiv \mathsf{M}^{-1} \cdot \mathbf{v}$.

A.2.2.4 *Special case I: Real Hermitian Matrix*

A real matrix is said to be *Hermitian* if its transpose, denoted $M^\top$ (i.e., $M_{ij}^\top = M_{ji}$), satisfies the identity $M^\top = M$, which requires that $c = b$ in Eq. (A.20). In this case, the eigenvalues are automatically real

$$\lambda_\pm = \left(\frac{a+d}{2}\right) \pm \sqrt{\left(\frac{a-d}{2}\right)^2 + b^2},$$

and the associated eigenvectors (A.26), which are defined with

$$\mu_\pm = -\left(\frac{a-d}{2b}\right) \pm \sqrt{1 + \left(\frac{a-d}{2b}\right)^2},$$

are automatically orthogonal to each other ($\mathbf{e}_+ \cdot \mathbf{e}_- = 0$) since $\mu_+ \mu_- \equiv -1$.

A.2.2.5 *Special case II: Rotation Matrix*

Another special matrix is given by the rotation matrix

$$R = \begin{pmatrix} \cos\theta & \sin\theta \\ -\sin\theta & \cos\theta \end{pmatrix}, \tag{A.32}$$

with determinant $\det(R) = 1$ and trace $\mathrm{Tr}(R) = 2\cos\theta = \exp(i\theta) + \exp(-i\theta)$. The rotation matrix (A.32) is said to be *unitary* since its transverse $R^\top$ is equal to its inverse $R^{-1} = R^\top$ (which is possible only if its determinant is one).

The eigenvalues of the rotation matrix (A.32) are $\exp(\pm i\theta)$ and the eigenvectors are

$$\mathbf{e}_\pm = \begin{pmatrix} 1 \\ \pm i \end{pmatrix}.$$

Note that the rotation matrix (A.32) can be written as $R = \exp(i\theta\,\boldsymbol{\sigma})$, where the matrix

$$\boldsymbol{\sigma} = \begin{pmatrix} 0 & -i \\ i & 0 \end{pmatrix}$$

(also known as the Pauli *spin* matrix σ_2) satisfies the properties $\sigma^{2n} = I$ and $\sigma^{2n+1} = \boldsymbol{\sigma}$ and, thus, we find

$$\exp(i\theta\,\boldsymbol{\sigma}) = \sum_{n=0}^{\infty} \frac{(i\theta)^n}{n!}\,\sigma^n = \cos\theta\,I + i\sin\theta\,\boldsymbol{\sigma} = R.$$

Note that the time derivative of the rotation matrix (A.32) satisfies the property

$$R^{-1} \cdot \dot{R} = i\dot{\theta}\,\boldsymbol{\sigma} = \begin{pmatrix} 0 & \dot{\theta} \\ -\dot{\theta} & 0 \end{pmatrix}.$$

Lastly, we note that the rotation matrix (A.32) can be used to *diagonalize* a real Hermitian matrix

$$M = \begin{pmatrix} a & b \\ b & d \end{pmatrix},$$

by constructing the new matrix

$$\overline{M} = R^\top \cdot M \cdot R = \begin{pmatrix} \overline{a} & \overline{b} \\ \overline{b} & \overline{d} \end{pmatrix}, \qquad (A.33)$$

where

$$\overline{a} = a + (d - a) \sin^2 \theta - b \sin 2\theta,$$

$$\overline{b} = b \cos 2\theta - \frac{1}{2}(d - a) \sin 2\theta,$$

$$\overline{d} = d - (d - a) \sin^2 \theta + b \sin 2\theta.$$

Next, by setting the non-diagonal element $\overline{b} \equiv 0$ (assuming that $d > a$), we obtain

$$\tan 2\theta = \frac{2b}{d - a} \quad \rightarrow \quad \begin{cases} \cos 2\theta = (d - a)/\sqrt{\sigma^2 - 4\Delta} \\ \\ \sin 2\theta = 2b/\sqrt{\sigma^2 - 4\Delta} \end{cases},$$

where $\sigma = a + d$ and $\Delta = ad - b^2$ denote the trace and determinant of M, respectively. Hence $\overline{M}$ becomes a diagonal matrix

$$\overline{M} = \begin{pmatrix} \lambda_- & 0 \\ 0 & \lambda_+ \end{pmatrix}, \qquad (A.34)$$

where the diagonal components are

$$\lambda_\pm = \frac{\sigma}{2} \pm \frac{1}{2}\sqrt{\sigma^2 - 4\Delta}.$$

Note that, since $\overline{a} + \overline{d} = a + d = \lambda_+ + \lambda_-$ and $\overline{a}\overline{d} - \overline{b}^2 = ad - b^2 = \lambda_+ + \lambda_-$, the trace and determinant of $\overline{M}$ are the same as that of M, i.e., the trace and determinant of any real Hermitian matrix are invariant under the *congruence* transformation (A.33).

A.3   Numerical Analysis

The nonlinear ordinary differential equations obtained in this course are often impossible to solve analytically in terms of known mathematical functions. Since numerical software is often readily available to students (e.g.,

Mathematica, Maple, or Mathlab), translating physical equations into dimensionless equations is a useful skill to acquire.

For example, the physical equation for the pendulum

$$\ddot{\theta} + \omega_\mathrm{g}^2 \sin\theta = 0$$

can be translated into the dimensionless equation

$$\theta''(\tau) + \sin\theta(\tau) = 0,$$

where $\omega_\mathrm{g} = \sqrt{g/\ell}$ and $\theta(\tau)$ is a function of the dimensionless time $\tau = \omega_\mathrm{g} t$. The great advantage of this dimensionless formulation is that the pendulum problem can be solved for all possible values of ω_g. The dimensionless pendulum can thus be solved by using the initial conditions $\theta(0)$ and $\theta'(0)$ determined from the dimensionless energy equation

$$\epsilon = (\theta')^2/2 + (1 - \cos\theta).$$

Hence, by choosing ϵ and the initial angle θ_0, we can determine the initial velocity $\theta_0' \equiv \pm\sqrt{2[\epsilon - (1 - \cos\theta_0)]}$.

Note that it is often preferable to adopt a Hamiltonian representation when numerically integrating equations of motion. This means that, instead of solving k second-order ordinary differential equations (ODEs), we are solving $2k$ first-order ODEs. For example, for the pendulum problem, we numerically solve Hamilton's equations $\theta'(\tau) = p(\tau)$ and $p'(\tau) = -\sin\theta(\tau)$, which allows us to easily plot the orbits of the pendulum in terms of the phase-space coordinates (θ, p).

When we consider nonlinear coupled equations such as Eqs. (2.47) and (2.48), which describe the motion on the surface of an inverted cone of apex angle α. It is desirable to choose a *clock* frequency needed to define a dimensionless time. By introducing $\omega_\mathrm{g} = \sqrt{(g/s_0)}\cos\alpha$ and $\tau = \omega_\mathrm{g} t$, we obtain the dimensionless equations

$$\sigma'' = -1 + \frac{1}{\sigma^3} \quad \text{and} \quad \theta' = \frac{1}{\sigma^2 \sin\alpha},$$

where $\sigma \equiv s/s_0$ becomes the distance normalized to

$$s_0 = \left(\frac{p_\theta^2}{m^2 g \sin^2\alpha \cos\alpha}\right)^{\frac{1}{3}}.$$

Note that, while several (physical) parameters appear in Eq. (2.48), the normalized equation $\sigma'' + 1 = \sigma^{-3}$ contains no dimensionless parameters at all, while the equation for θ' only requires that the cone angle α (which can even be absorbed in a new definition of θ).

The following sample Mathematica code (v 4.0) was written to generate solutions of the problem of constrained motion on the surface of a cone and to create Fig. (2.9).

```
s₀ = 3
p₀ = 0
q₀ = 0

a = π/8

t₀ = 0.0
t₁ = 20.0

solution1 = NDSolve[{
    s'[t]  ==  p[t],
    p'[t]  ==  -1 + 1/s[t]³,
    q'[t]  ==  1/(Sin[a] s[t]²),
    s[0]   ==  s₀,
    p[0]   ==  p₀,
    q[0]   ==  q₀ },
    {s[t], p[t], q[t]}, {t, t₀, t₁}]

plot1 = ParametricPlot[Evaluate[{s[t] Sin[a] Cos[q[t]], s[t] Sin[a] Sin[q[t]]}
    /. solution1], {t, t₀, t₁}, AxesLabel → {"x-axis", "y-axis"}]

plot2 = ParametricPlot[Evaluate[{s[t] Sin[a] Cos[q[t]], s[t] Cos[a]}
    /. solution1], {t, t₀, t₁}, AxesLabel → {"x-axis", "z-axis"}]
```

Note here that we are numerically solving 3 first-order ODEs: $\sigma' = p$, $p' = -1 + 1/\sigma^3$, and $\theta' = 1/(\sigma^2 \sin\alpha)$. The command plot1 generates the "top view" of Fig. 2.9 while plot2 generates the "side view". If we need to change the parameters or initial conditions for our numerical solution, we can create a new solution #, e.g., solution 2 = NDSolve[{···}].

Elliptic Functions and Integrals*

The Jacobi and Weierstrass elliptic functions used to be part of the standard mathematical arsenal of physics students [17]. They appear as solutions of many important problems in classical mechanics: the motion of a planar pendulum (Jacobi), the motion of a force-free asymmetric top (Jacobi), the motion of a spherical pendulum (Weierstrass), and the motion of a heavy symmetric top with one fixed point (Weierstrass). The problem of the planar pendulum, in fact, can be used to construct the general connection between the Jacobi and Weierstrass elliptic functions. The easy access to mathematical software by physics students suggests that they might reappear as useful tools in the undergraduate curriculum.[1]

B.1   Jacobi Elliptic Functions

B.1.1   *Definitions and Notation*

We begin our introduction of elliptic functions with the more familiar Jacobi elliptic functions [14]. The Jacobi elliptic function $\text{sn}(z \,|\, m)$ is defined in terms of the inverse-function formula

$$
\begin{aligned}
z &= \int_0^\varphi \frac{d\theta}{\sqrt{1 - m \sin^2 \theta}} \\
&= \int_0^{\sin \varphi} \frac{dy}{\sqrt{(1 - y^2)\,(1 - m\,y^2)}} \\
&\equiv \text{sn}^{-1}(\sin \varphi \,|\, m),
\end{aligned}
\tag{B.1}
$$

where the modulus m is a positive number and the amplitude φ varies from 0 to 2π. From this definition, we easily check that $\text{sn}^{-1}(\sin \varphi \,|\, 0) =$

---

[1] An extended version of this Appendix can be found at http://arxiv.org/abs/0711.4064.

$\sin^{-1}(\sin\varphi) = \varphi$. The solution to the differential equation

$$\left(\frac{dy}{dz}\right)^2 = (1-y^2)\,(1-m\,y^2) \tag{B.2}$$

is expressed in terms of the Jacobi elliptic function

$$y(z) = \begin{cases} \text{sn}(z|m) & \text{(for } m < 1\text{)}, \\[2mm] m^{-1/2}\,\text{sn}\left(m^{1/2}\,z\mid m^{-1}\right) & \text{(for } m > 1\text{)}. \end{cases} \tag{B.3}$$

By using the transformation $y = \sin\varphi$, the Jacobi differential equation (B.2) is also written as

$$\left(\frac{d\varphi}{dz}\right)^2 = 1 - m\,\sin^2\varphi, \tag{B.4}$$

and the solution to this equation is $\varphi(z) = \sin^{-1}[\text{sn}(z|m)]$ for $m < 1$.

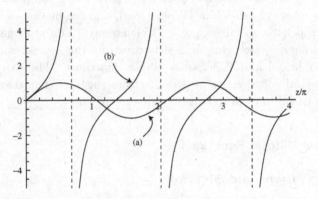

Fig. B.1   Plots of (a) $\text{sn}(z|m)$ and (b) $-i\,\text{sn}(iz|m)$ for $m = 1/16$ showing the real and imaginary periods $4\,K(m)$ and $4\,i\,K'(m)$.

The function $\text{sn}(z|m)$ has a purely-real period $4\,K$, where the quarter-period K is defined as

$$K \equiv K(m) = \int_0^{\pi/2} \frac{d\theta}{\sqrt{1 - m\,\sin^2\theta}} \tag{B.5}$$

and a purely-imaginary period $4\,iK'$, where the quarter-period K' is defined as (with the complementary modulus $m' \equiv 1 - m$)

$$i\,K' \equiv i\,K(m') = i\int_0^{\pi/2} \frac{d\theta}{\sqrt{1 - m'\,\sin^2\theta}}. \tag{B.6}$$

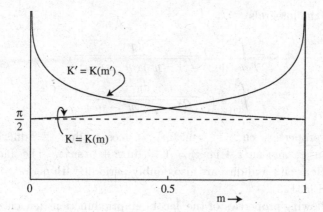

Fig. B.2   Plots of the quarter periods $K = K(m)$ and $K' = K(m') = K(1 - m)$.

Figure B.1 shows plots of $\operatorname{sn} z$ and $-i\operatorname{sn}(iz)$ for $m = 1/16$, which exhibit both a real period and an imaginary period. Note that, while the Jacobi elliptic function $\operatorname{sn} z$ alternates between -1 and $+1$ for real values of z (with zeroes at $2nK$), it also exhibits singularities for imaginary values of z at $(2n + 1)\,iK'$ $(n = 0, 1, ...)$. Furthermore, as $m \to 0$ (and $m' \to 1$), we find $K \to \pi/2$ (or $4K \to 2\pi$) and $|K'| \to \infty$ (see Fig. B.2), and so $\operatorname{sn} z \to \sin z$ becomes singly-periodic.

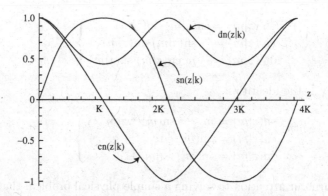

Fig. B.3   Plots of $\operatorname{sn}(z|m)$, $\operatorname{cn}(z|m)$, and $\operatorname{dn}(z|m)$ from $z = 0$ to $4K(m)$ for $m = 0.81$.

The additional Jacobi elliptic functions $\operatorname{cn}(z\,|\,m)$ and $\operatorname{dn}(z\,|\,m)$ are defined from the integrals

$$z = \int_{cn(z|m)}^{1} \frac{dy}{\sqrt{(1-y^2)\,(m'+m\,y^2)}}, \tag{B.7}$$

$$= \int_{dn(z|m)}^{1} \frac{dy}{\sqrt{(1-y^2)\,(y^2-m')}}, \tag{B.8}$$

with the properties $\text{cn}\,z \equiv \text{cn}(z|m) = \cos\varphi$, $\text{dn}\,z \equiv \text{dn}(z|m) = \sqrt{1-m\sin^2\varphi}$, and $\text{sn}^2 z + \text{cn}^2 z = 1 = \text{dn}^2 z + m\,\text{sn}^2 z$. The Jacobi elliptic functions $\text{cn}\,z$ and $\text{dn}\,z$ are also doubly-periodic with periods $4\,K$ and $4i\,K'$ (see Fig. B.3).

The following properties of the Jacobi elliptic functions $(\text{sn}, \text{cn}, \text{dn})$ are useful. First, we find the limits:

$$\begin{pmatrix} \text{sn}(z|0) \\ \text{cn}(z|0) \\ \text{dn}(z|0) \end{pmatrix} = \begin{pmatrix} \sin z \\ \cos z \\ 1 \end{pmatrix} \tag{B.9}$$

and

$$\begin{pmatrix} \text{sn}(z|1) \\ \text{cn}(z|1) \\ \text{dn}(z|1) \end{pmatrix} = \begin{pmatrix} \tanh z \\ \text{sech}\,z \\ \text{sech}\,z \end{pmatrix}. \tag{B.10}$$

Next, we find the derivatives with respect to the argument z:

$$\left. \begin{array}{l} \text{sn}'(z|m) = \text{cn}(z|m)\,\text{dn}(z|m) \\ \text{cn}'(z|m) = -\,\text{sn}(z|m)\,\text{dn}(z|m) \\ \text{dn}'(z|m) = -\,m\,\text{cn}(z|m)\,\text{sn}(z|m) \end{array} \right\}, \tag{B.11}$$

and, if $m > 1$, the identities:

$$\left. \begin{array}{l} \text{sn}(z|m) = m^{-1/2}\,\text{sn}(m^{1/2}\,z|m^{-1}) \\ \text{cn}(z|m) = \text{dn}(m^{1/2}\,z|m^{-1}) \\ \text{dn}(z|m) = \text{cn}(m^{1/2}\,z|m^{-1}) \end{array} \right\}. \tag{B.12}$$

We now turn our attention to solving a simple physical problem that highlights the periodic properties the Jacobi elliptic functions (B.1) and (B.7)-(B.8). Already, we have seen how the problems of the planar pendulum in Sec. 3.5.3 and the force-free asymmetric top in Sec. 7.2.3 can be solved simply and explicitly in terms of the Jacobi elliptic functions $(\text{sn}, \text{cn}, \text{dn})$.

B.1.2 Motion in a Quartic Potential

We look at particle orbits in the (dimensionless) quartic potential $U(x) = 1 - x^2/2 + x^4/16$. Here, the turning points for $E \equiv e^2 = U(x)$ are

$$
\left. \begin{array}{c}
\pm 2\sqrt{1+e} \quad (\text{for } e > 1) \\[2mm]
0 \text{ and } \pm\sqrt{8} \ (\text{for } e = 1) \\[2mm]
\pm 2\sqrt{1 \pm e} \quad (\text{for } e < 1)
\end{array} \right\} . \tag{B.13}
$$

Each orbit is solved using the initial condition $x_0 = 2\sqrt{1+e}$ with the initial velocity $\dot{x}_0 < 0$:

$$
\begin{aligned}
t(x) &= -\int_{2\sqrt{1+e}}^{x} \frac{dy}{\sqrt{2(e^2-1) + y^2(1 - y^2/8)}} \\
&= -\int_{2\sqrt{1+e}}^{x} \frac{\sqrt{8}\, dy}{\sqrt{[4(e+1) - y^2][y^2 + 4(e-1)]}} \\
&= \frac{1}{\sqrt{e}} \int_{0}^{\Phi(x)} \frac{d\varphi}{\sqrt{1 - m\sin^2\varphi}},
\end{aligned} \tag{B.14}
$$

where $m \equiv (1+e)/2e$ while we used the trigonometric substitution $y = 2\sqrt{1+e}\cos\varphi$ with

$$
\Phi(x) \equiv \cos^{-1}\left[\frac{x}{2\sqrt{1+e}}\right] \tag{B.15}
$$

to obtain the last expression in Eq. (B.14). The Jacobi elliptic solutions obtained from Eq. (B.14) are shown in Fig. B.4 for the orbit (a), with $e > 1$, the separatrix orbit (b), with $e = 1$, and the orbit (c), with $e < 1$.

For $e > 1$ (i.e., $m < 1$), corresponding to orbit (a) in Fig. B.4, we use Eq. (B.1) to find

$$
\sin\Phi(x) = \text{sn}(\sqrt{e}\,t|m) = \sqrt{1 - \frac{x^2(t)}{4(1+e)}},
$$

which yields the phase-portrait coordinates $(x, \dot{x})$:

$$
\left. \begin{array}{l}
x(t) = 2\sqrt{1+e}\ \text{cn}(\sqrt{e}\,t|m) \\[3mm]
\dot{x}(t) = -2\sqrt{e(1+e)}\ \text{sn}(\sqrt{e}\,t|m)\,\text{dn}(\sqrt{e}\,t|m)
\end{array} \right\} , \tag{B.16}
$$

where the velocity $\dot{x}(t)$ is obtained by using Eq. (B.11). For $e = 1$ (i.e., the separatrix orbit with $m = 1$), corresponding to orbit (b) in Fig. B.4, the phase-portrait coordinates become

$$
\left. \begin{array}{l}
x(t) = \sqrt{8}\ \text{sech}\,t \\[3mm]
\dot{x}(t) = -\sqrt{8}\ \text{sech}\,t\,\tanh t
\end{array} \right\} , \tag{B.17}
$$

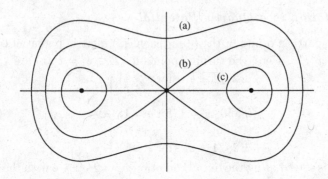

Fig. B.4   Phase portrait for orbits (B.16)-(B.18) of the quartic potential $U(x) = 1 - x^2/2 + x^4/16$ for (a) e > 1, (b) e = 1 (separatrix), and (c) e < 1.

where the limits (B.10) were applied to Eq. (B.16). Lastly, for e < 1 (i.e., $m > 1$), corresponding to orbit (c) in Fig. B.4, we apply the relations (B.12) on Eq. (B.16) to obtain

$$\left.\begin{array}{l} x(t) = 2\sqrt{1+e}\ \operatorname{dn}(\tau\,|m^{-1}) \\[2mm] \dot{x}(t) = -\sqrt{8}\,e\ \operatorname{sn}(\tau\,|m^{-1})\,\operatorname{cn}(\tau\,|m^{-1}) \end{array}\right\}, \qquad (B.18)$$

where $\tau = t\,\sqrt{(1+e)/2}$. The orbits (B.16)-(B.18) are combined to yield the phase portrait for the quartic potential shown in Fig. B.4.

B.2   Weierstrass Elliptic Functions

B.2.1   *Definitions and Notation*

The Weierstrass elliptic function $\wp(z; g_2, g_3)$ is defined as the solution of the differential equation [16]

$$(ds/dz)^2 = 4\,s^3 - g_2\,s - g_3$$
$$\equiv 4\,(s - e_1)\,(s - e_2)\,(s - e_3). \qquad (B.19)$$

Here, (e_1, e_2, e_3) denote the roots of the cubic polynomial $4s^3 - g_2\,s - g_3$ (such that $e_1 + e_2 + e_3 = 0$), where the invariants g_2 and g_3 are defined in terms of the cubic roots as

$$\left.\begin{array}{l} g_2 = -4\,(e_1\,e_2 + e_2\,e_3 + e_3\,e_1) = 2\,(e_1^2 + e_2^2 + e_3^2) \\[2mm] g_3 = 4\,e_1\,e_2\,e_3 \end{array}\right\}, \qquad (B.20)$$

and $\Delta = g_2^3 - 27 g_3^2$ is the modular discriminant. Since physical values for the constants g_2 and g_3 are always real (and $g_2 > 0$), then either all three roots are real or one root (say e_a) is real and we have a conjugate pair of complex roots (e_b, e_b^*) with $\text{Re}(e_b) = -e_a/2$. The applications of Weierstrass elliptic functions are analyzed in terms of four different cases based on the signs of $(g_3, \Delta) = [(-,-), (-,+), (+,-), (+,+)]$, with two special cases $(g_3 \neq 0, \Delta = 0)$ and $(g_3 = 0, \Delta > 0)$.

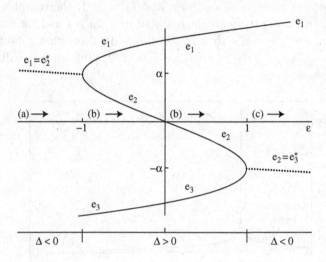

Fig. B.5   Cubic roots (e_1, e_2, e_3) as a function of $\epsilon \equiv (3/g_2)^{3/2} g_3$ with fixed value g_2, where $\alpha \equiv \sqrt{g_2/12}$ and $\Delta = g_2^3 (1 - \epsilon^2)$. The three roots satisfy $e_1 + e_2 + e_3 = 0$.

In general, the roots (e_1, e_2, e_3) of the cubic polynomial on the right side of Eq. (B.19) can be expressed in terms of the parameters $\alpha \equiv \sqrt{g_2/12}$ and $\epsilon \equiv (3/g_2)^{3/2} g_3 \equiv -\cos\varphi$ as

$$\begin{pmatrix} e_1 \\ e_2 \\ e_3 \end{pmatrix} \equiv 2\alpha \begin{pmatrix} \cos[(\varphi - \pi)/3] \\ \cos[(\varphi + \pi)/3] \\ -\cos(\varphi/3) \end{pmatrix}, \tag{B.21}$$

and the discriminant is

$$\Delta = g_2^2 - 27 g_3^2 = g_2^3 (1 - \epsilon^2). \tag{B.22}$$

These cubic roots are shown in Fig. B.5 as a function of ϵ for $\alpha = 1/2$ (i.e., $g_2 = 3$); the polynomial $4s^3 - g_2 s - g_3$ is positive (and ds/dz is real) to the left of the curve and negative (and ds/dz is imaginary) to the right of the curve. The three cubic roots (as shown in Fig. B.5) are connected smoothly

in the complex φ-plane

$$\varphi = \begin{cases} -i\,\psi & (\psi \geq 0,\ \epsilon \leq -1) \\ \phi & (0 \leq \phi \leq \pi,\ -1 \leq \epsilon \leq 1) \\ \pi + i\,\psi & (\psi \geq 0,\ \epsilon \geq 1) \end{cases} \tag{B.23}$$

Here, for $\epsilon \leq -1$, the imaginary phase $\varphi \equiv -i\,\psi$ (with $\psi \geq 0$) yields the complex-conjugate roots $e_1 = a - ib = e_2^*$ (with $b > 0$) and the real root $e_3 = -2a < -1$; for $-1 \leq \epsilon \leq 1$, the real phase $\varphi \equiv \phi$ (with $0 \leq \phi \leq \pi$) yields three real roots $e_1 > e_2 > e_3$; and for $\epsilon \geq 1$, the complex phase $\varphi = \pi + i\,\psi$ (with $\psi \geq 0$) yields the real root $e_1 = 2a > 1$ and the complex-conjugate roots $e_2 = -a - ib = e_3^*$ (with $b > 0$). Note that $e_2 = 0$ (and $e_1 = \sqrt{3}\,\alpha = -e_3$) for $g_3 = 0$ (i.e., $\phi = \pi/2$); this case is called the *lemniscatic* case.

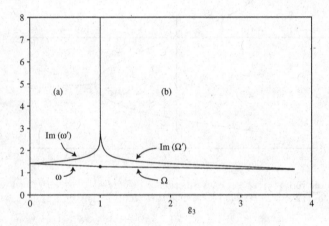

Fig. B.6 Plots of (a) ω and $\mathrm{Im}(\omega')$ for $0 < g_3 < 1$ and $g_2 = 3$ (i.e., $\Delta > 0$) and (b) Ω and $\mathrm{Im}(\Omega')$ for $g_3 > 1$ and $g_2 = 3$ (i.e., $\Delta < 0$). Note that $\omega'(g_2, 0) = i\omega(g_2, 0)$, $\Omega(g_2, 1) = \omega(g_2, 1)$, and both (Ω, Ω') decrease to zero as g_3 becomes infinite.

For $0 < \epsilon < 1$ (i.e., $\Delta > 0$), $\wp(z)$ has two different periods $2\,\omega$ and $2\,\omega'$ along the real and imaginary axes, respectively, with the half-periods ω and ω' defined as

$$\omega(g_2, g_3) = \int_{e_1}^{\infty} \frac{ds}{\sqrt{4s^3 - g_2\,s - g_3}}, \tag{B.24}$$

$$\omega'(g_2, g_3) = i \int_{-\infty}^{e_3} \frac{ds}{\sqrt{|4s^3 - g_2\,s - g_3|}}. \tag{B.25}$$

The plots of $\omega(g_2, g_3)$ and $\omega'(g_2, g_3)$ are shown in Fig. B.6 for $g_2 = 3$ as functions of g_3. Note that for $g_3 = 0$ (with $e_1 = -e_3$ and $e_2 = 0$), we find that $\omega' \equiv i\omega$ while $|\omega'|$ approaches infinity as g_3 approaches one.

Table B.1   Cubic roots (e_1, e_2, e_3) and half periods $(\omega_1, \omega_2, \omega_3)$ for $g_2 = 3$ $(\alpha = 1/2)$.

(g_3, Δ)	e_1	e_2	e_3	ω_1	ω_2	ω_3
$(-,-)$	$a - ib$	$a + ib$	$-2a < -1$	$\|\Omega'\| + i\Omega/2$	$-\|\Omega'\| + i\Omega/2$	$-i\Omega$
$(-,+)$	$d > 0$	$c - d > 0$	$-c < 0$	$\|\omega'\|$	$i\omega - \|\omega'\|$	$-i\omega$
$(+,+)$	$c > 0$	$d - c > 0$	$-d < 0$	ω	$-\omega - \omega'$	ω'
$(+,-)$	$2a > 1$	$-a - ib$	$-a + ib$	Ω	$-\Omega/2 - \Omega'$	$-\Omega/2 + \Omega'$

For $\epsilon > 1$ (i.e., $\Delta < 0$), on the other hand, $\wp(z)$ has two different periods $2\,\Omega$ and $2\,\Omega'$ along the real and imaginary axes, respectively, with the half-periods Ω and Ω' defined as

$$\Omega(g_2, g_3) = \int_{e_1}^{\infty} \frac{ds}{\sqrt{4s^3 - g_2\, s - g_3}}, \tag{B.26}$$

$$\Omega'(g_2, g_3) = i \int_{-\infty}^{e_1} \frac{ds}{\sqrt{|4s^3 - g_2\, s - g_3|}}. \tag{B.27}$$

The plots of $\Omega(g_2, g_3)$ and $\Omega'(g_2, g_3)$ are shown in Fig. B.6 for $g_2 = 3$ as functions of g_3. Note that $\omega(g_2, 1) = \Omega(g_2, 1)$, $|\Omega'|$ approaches infinity as g_3 approaches one, and that both Ω and Ω' approach zero as g_3 approaches infinity.

Table B.1 shows the cubic roots $e_i = (e_1, e_2, e_3)$, defined by Eq. (B.21), and the half periods $\omega_i = (\omega_1, \omega_2, \omega_3)$, defined as

$$\omega_i(g_2, g_3) \equiv \int_{e_i}^{\infty} \frac{ds}{\sqrt{4s^3 - g_2\, s - g_3}}$$

$$= \int_{e_i}^{\infty} \frac{ds}{2\sqrt{(s - e_1)(s - e_2)(s - e_3)}}. \tag{B.28}$$

The cubic roots and half periods satisfy the following properties:

$$\left.\begin{aligned} \wp(\omega_i) &= e_i \\ \wp(z + \omega_i) &= e_i + (e_i - e_j)(e_i - e_k)[\wp(z) - e_i]^{-1} \\ \wp(z + 2\omega_i) &= \wp(z) \end{aligned}\right\}, \tag{B.29}$$

where $i \neq j \neq k$ so that $\wp(\omega_i + \omega_j) = e_k$. Figure B.7 shows the plots of $\wp(z + \omega_2)$ and $\wp(z + \omega_3)$ for one complete period from $z = 0$ to $2\,\omega_1$, which clearly satisfies the identities (B.29).

The Weierstrass elliptic function $\wp(z; g_2, g_3)$ obeys the homogeneity relation

$$\wp\left(\lambda z; \lambda^{-4} g_2, \lambda^{-6} g_3\right) = \lambda^{-2}\, \wp(z; g_2, g_3), \tag{B.30}$$

where $\lambda \neq 0$. By choosing $\lambda = -1$, for example, we readily verify that the Weierstrass elliptic function has even parity, i.e., $\wp(-z; g_2, g_3) =$

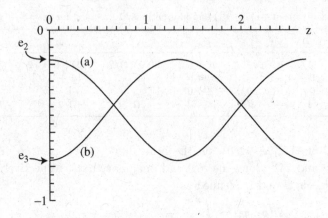

Fig. B.7   Plots of (a) $\wp(z+\omega_2)$ and (b) $\wp(z+\omega_3)$ for $g_2 = 3$ and $g_3 = 0.5$ (with $\epsilon = 0.5$) over one complete period from 0 to $2\omega_1$. Note that $\wp(\omega_j) = e_j$ for $j = 2$ or 3 and $\wp(\omega_i + \omega_j) = e_k$, for $i = 1$ and $(j, k) = (2, 3)$ or $(3, 2)$.

$\wp(z; g_2, g_3)$. On the other hand, for $\lambda = i$, we find that the half-period assignments for $g_3 < 0$ in Table B.1 are based on the relation

$$\wp(z; g_2, g_3) = -\wp(iz; g_2, |g_3|). \qquad (B.31)$$

For example, for $-1 < g_3 < 0$ (and $\Delta > 0$), we find for $\wp(\omega_1; g_2, g_3)$:

$$\wp(|\omega'|; g_2, -|g_3|) = -\wp(\omega'; g_2, |g_3|) = -(-d) = d,$$

which corresponds exactly to $e_1 = d$ found in Table B.1 for the case $(g_3, \Delta) = (-, +)$.[2]

In general, the connections between the half-periods

$$(\omega_1^+, \omega_2^+, \omega_3^+) \rightarrow (\omega_1^-, \omega_2^-, \omega_3^-)$$

and the cubic roots $(e_1^+, e_2^+, e_3^+) \rightarrow (e_1^-, e_2^-, e_3^-)$ as g_3 changes sign from positive $(+)$ to negative $(-)$ are found in Table B.1 to be

$$\left.\begin{array}{c} (\omega_1^-, \omega_2^-, \omega_3^-) \equiv (-i\,\omega_3^+, -i\,\omega_2^+, -i\,\omega_1^+) \\[2mm] (e_1^-, e_2^-, e_3^-) \equiv (-e_3^+, -e_2^+, -e_1^+) \end{array}\right\}. \qquad (B.32)$$

Once again, these connections follow a non-standard convention. For example, according to the standard convention [16] for the case $(g_3, \Delta) = (+, -)$,

---

[2]The reader should be warned that only the case $(g_3, \Delta) = (+, +)$ in Table B.1 follows the standard mathematical convention [16]. The convention for the remaining cases $(g_3, \Delta) = [(-, -), (-, +), (+, -)]$ in Table B.1 are based on the output of Mathematica, on which Eq. (B.21) and the Weierstrass path (B.23) are based, and the convention adopted for ω_2 satisfies the condition $\omega_1 + \omega_2 + \omega_3 = 0$.

the root e_2 is real (while $e_1^* = e_3$) and the corresponding half-period ω_2 is also real (in contrast to the convention adopted in Table B.1). The connections (B.32) shown in Table B.1 are simply based on the smooth dependence of the cubic roots on the single parameter ϵ (for fixed g_2). These connections enable us to describe consistent orbital dynamics in several problems in classical mechanics.

B.2.2   *Motion in a Cubic Potential*

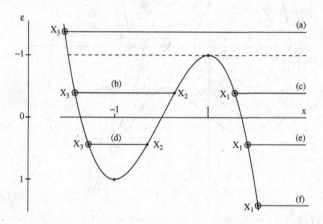

Fig. B.8   Cubic-potential energy levels $E = x - x^3/3$ showing orbits (a) $E > 2/3$ (unbounded orbits; $\epsilon < -1$), (b) and (c) $0 < E < 2/3$ (bounded and unbounded orbits; $-1 < \epsilon < 0$), (d) and (e) $-2/3 < E < 0$ (bounded and unbounded orbits; $0 < \epsilon < 1$), and (f) $E \leq -2/3$ (unbounded orbit; $\epsilon \geq 1$).

We have already seen that the solution of the sleeping top problem can be written in terms of the Weierstrass elliptic function (Sec. 8.1.1). As an additional physical problem, we consider particle orbits in a (dimensionless) cubic potential $U(x) = x - x^3/3$. Here, the cubic-potential orbits $x(t)$ are solutions of the differential equation

$$\dot{x}^2 = 2\left(E - x + \frac{x^3}{3} \right)$$

$$\equiv \frac{2}{3}\,(x - x_1)\,(x - x_2)\,(x - x_3), \tag{B.33}$$

and the turning points (x_1, x_2, x_3) are shown in Fig. B.8 (with $x_1 + x_2 + x_3 = 0$). By writing $x(t) = 6\,s(t)$, Eq. (B.33) is transformed into the standard Weierstrass elliptic equation (B.19), where the invariants are $g_2 = 1/3$ and

$g_3 = -E/18$, so that $\epsilon \equiv -3E/2$. Note that bounded orbits exist only for $-1 < \epsilon < 1$ (i.e., $\Delta > 0$).

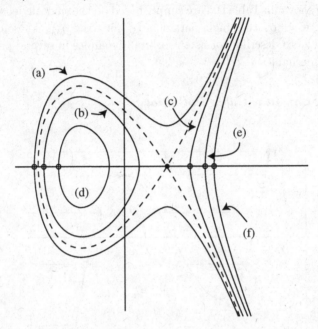

Fig. B.9 Plots of $\dot{x}(t)$ versus $x(t)$ for cubic potential (B.33) shows bounded and unbounded orbits: Orbit (a) $E > 2/3$ ($\epsilon < -1$ and $\Delta < 0$); orbits (b)-(e) $-2/3 < E < 2/3$ ($-1 < \epsilon < 1$ and $\Delta > 0$); and orbit (f) $E < -2/3$ ($\epsilon > 1$ and $\Delta < 0$). The dotted lines are the bounded and unbounded separatrix orbits for $E = 2/3$ and circles denote particle positions at $t = 0$.

The cubic-potential solution for Eq. (B.33) is

$$x(t) = 6 \, \wp(t + \gamma), \tag{B.34}$$

where the constant γ is determined from the initial condition $x(0)$. Figure B.9 shows the orbits (a)-(f) associated with initial conditions identified by a circle and a qualitative description of these orbits is summarized in Table B.2. Note that the turning points $x_i = 6 \, e_i$ ($i = 1, 2, 3$) are simply related to the standard cubic roots e_i. Lastly, the separatrix solution is obtained from orbit (b) as E approaches $2/3$ and the period $2 \, |\omega'|$ becomes infinite.

Lastly, we note that the imaginary time range for orbit (a) takes into account the relation (B.31) since $g_3 < 0$ for this orbit. In addition, the connections (B.32) allow us to describe the orbits (a)-(f) in Figs. B.8 and B.9 (and Table B.2) smoothly as the single (energy) parameter ϵ is varied.

Table B.2 Bounded and unbounded orbits in a cubic potential (see Figs. B.8 and B.9).

Orbit	Energy	Time Range	Constant γ	Period
(a)	$E > 2/3$	$-i\Omega < t < i\Omega$	$-i\Omega$	Unbounded
(b)	$0 < E < 2/3$	$0 < t < 2\,|\omega'|$	$-i\omega$	$2\,|\omega'|$
(c)	$0 < E < 2/3$	$-|\omega'| < t < |\omega'|$	$|\omega'|$	Unbounded
(d)	$-2/3 < E < 0$	$0 < t < 2\omega$	ω'	2ω
(e)	$-2/3 < E < 0$	$-\omega < t < \omega$	ω	Unbounded
(f)	$E < -2/3$	$-\Omega < t < \Omega$	Ω	Unbounded

B.3 Connection between Elliptic Functions

In this Section, we return to the planar pendulum problem of Sec. 3.5.3 to establish a connection between the Jacobi and Weierstrass elliptic functions. First, we write $z = 1 - \cos\theta$ (i.e., $0 < z < 2$) and transform Eq. (3.34) into the cubic-potential equation

$$(z')^2 = 2z(2-z)(\epsilon - z),\qquad(B.35)$$

with roots at $z = 0, 2$ and $\epsilon \equiv E/(mg\ell)$. When $\epsilon < 2$, the motion is periodic between $z = 0$ and $z = \epsilon$, while the motion is periodic between $z = 0$ and $z = 2$ for $\epsilon > 2$. We recover the standard Weierstrass differential equation (B.19) by setting

$$z(\tau) = 2\wp(\tau + \gamma) + \mu,\qquad(B.36)$$

where $\mu \equiv (\epsilon + 2)/3$ and the constant γ is determined from the initial condition $z(0)$.

The root corresponding to $z = 0$ is labeled $e_c = -\mu/2$, the root corresponding to $z = 2$ is labeled $e_b = 1 - \mu/2$, and the root corresponding to $z = \epsilon$ is labeled $e_a = \mu - 1$ and we easily verify that $e_a + e_b + e_c = 0$ (see Fig. B.10). The Weierstrass invariants are $g_2 = 1 + 3(\mu - 1)^2$ and $g_3 = \mu(\mu - 1)(\mu - 2)$, and the modular discriminant is $\Delta = \epsilon^2(2 - \epsilon)^2 \geq 0$.

The planar pendulum is now discussed in terms of 4 cases labeled (a)-(d) in Fig. B.10. For cases (a) and (b), where $2/3 < \mu < 4/3$ (i.e., $0 < \epsilon < 2$), we find

$$e_3 = -\mu/2 < e_2 = \mu - 1 < e_1 = 1 - \mu/2,$$

so that

$$\left.\begin{aligned}\kappa &= (e_1 - e_3)^{1/2} = 1 \\ m &= (e_2 - e_3)/(e_1 - e_3) = (3\mu - 2)/2 = \epsilon/2 < 1\end{aligned}\right\}.\qquad(B.37)$$

For cases (c) and (d), where $\mu > 4/3$ (i.e., $\epsilon > 2$), we find

$$e_3 = -\mu/2 < e_2 = 1 - \mu/2 < e_1 = \mu - 1,$$

so that

$$\left.\begin{array}{l} \kappa = (e_1 - e_3)^{1/2} = (\epsilon/2)^{1/2} > 1 \\[2mm] m = (e_2 - e_3)/(e_1 - e_3) = 2/\epsilon < 1 \end{array}\right\}. \tag{B.38}$$

Figure B.11 shows a plot of g_3 as a function of the parameter ϵ, which can be used with the information presented in Table B.1 to describe the motion of the planar pendulum in terms of the Weierstrass elliptic function.

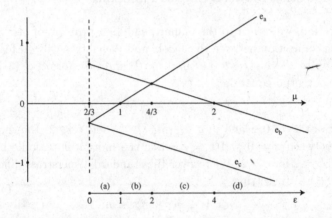

Fig. B.10 Plots of the cubic roots (e_a, e_b, e_c) as functions of $\mu = (\epsilon + 2)/3$. The cases (a)-(d) are discussed in the text. Note that for cases (a) and (b), or $\epsilon < 2$, we find $e_c < e_a < e_b$, while for cases (c) and (d), or $\epsilon > 2$ we find $e_c < e_b < e_a$. The bounded motion of the planar pendulum ($-1 \leq z \leq 1$) occurs between the two lowest cubic roots: $e_c < e_a$ (for $\epsilon < 2$) or $e_c < e_b$ (for $\epsilon > 2$).

We first consider case (a), where $0 < \epsilon < 1$ (i.e., $2/3 < \mu < 1$ and $g_3 > 0$), the periodic motion is bounded between $e_3 = -\mu/2$ (i.e., $z = 0$) and $e_2 = \mu - 1 < 0$ (i.e., $z = \epsilon$). Using the initial condition $z(0) = 0$, we find that $\wp(\gamma) = -\mu/2 \equiv e_3$ which implies that $\gamma = \omega'$ (see ω_3 in Table B.1 for $g_3 > 0$ and $\Delta > 0$). The Weierstrass solution of the planar pendulum for $0 < \epsilon < 1$ is therefore

$$z(\tau) = 2\,\wp(\tau + \omega') + \mu, \tag{B.39}$$

with the period of oscillation 2ω.

For case (b), where $1 < \epsilon < 2$ (i.e., $1 < \mu < 4/3$ and $g_3 < 0$), the periodic motion is bounded between $e_3 = -\mu/2$ (i.e., $z = 0$) and $e_2 = \mu - 1 < 0$ (i.e.,

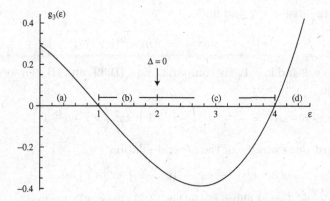

Fig. B.11 Plot of the Weierstrass invariant g_3 as a function of ϵ. For case (a), $g_3 > 0$ and $\Delta > 0$; for cases (b) and (c), $g_3 < 0$ and $\Delta \geq 0$; and for case (d) $g_3 > 0$ and $\Delta > 0$.

$z = \epsilon$). Using the initial condition $z(0) = 0$, we find that $\wp(\gamma) = -\mu/2 \equiv e_3$ which implies that $\gamma = -i\omega$ (see ω_3 in Table B.1 for $g_3 < 0$ and $\Delta \geq 0$). The Weierstrass solution of the planar pendulum for $1 < \epsilon < 2$ is

$$z(\tau) = 2\wp(\tau - i\omega) + \mu, \tag{B.40}$$

with the period of oscillation $2\,|\omega'|$. As expected, when $\epsilon \to 2$ (i.e., $\Delta \to 0$ and $m \to 1$), the period $2\,|\omega'|$ approaches infinity as we approach the separatrix.

For case (c), where $2 < \epsilon < 4$ (i.e., $4/3 < \mu < 2$ and $g_3 < 0$), the periodic motion is bounded between $e_3 = -\mu/2$ (i.e., $z = 0$) and $e_2 = 1 - \mu/2$ (i.e., $z = 2$), with the period of oscillation $2\,|\omega'|$. Using the initial condition $z(0) = 0$, we find that $\wp(\gamma) = -\mu/2 \equiv e_3$ which implies that $\gamma = -i\omega$ (see ω_3 in Table B.1 for $g_3 < 0$ and $\Delta \geq 0$) and thus the Weierstrass solution of the planar pendulum for $2 < \epsilon < 4$ is again given by Eq. (B.40). Note that the separatrix solution ($\epsilon = 2$) is represented by orbits (b) and (c) as $|\omega'| \to \infty$.

Lastly, for case (d), where $\epsilon > 4$ (i.e., $\mu > 2$ and $g_3 > 0$), the periodic motion is bounded between $e_3 = -\mu/2$ (i.e., $z = 0$) and $e_2 = 1 - \mu/2$ (i.e., $z = 2$). Using the initial condition $z(0) = 0$, we find that $\wp(\gamma) = -\mu/2 \equiv e_3$ which implies that $\gamma = \omega'$ (see ω_3 in Table B.1 for $g_3 > 0$ and $\Delta > 0$) and thus the Weierstrass solution of the planar pendulum for $\epsilon > 4$ is again given by Eq. (B.39).

We conclude our discussion of the planar pendulum by using the Jacobi and Weierstrass solutions of this problem to establish a general relation between these elliptic functions. First, we use the Jacobi elliptic solution

(3.39) for the case $\epsilon < 2$ and find

$$z(\tau) = 2 \sin^2 \frac{\theta}{2} = 2m \operatorname{sn}^2(\kappa \tau \mid m), \tag{B.41}$$

where $m = \epsilon/2$ and $\kappa = 1$. By comparing Eqs. (B.39) and (B.41), we obtain the relation

$$\wp(\tau + \omega_3) = -\frac{\mu}{2} + \left(\frac{3}{2}\mu - 1\right) \operatorname{sn}^2 \left(\tau \left| \frac{3}{2}\mu - 1\right.\right),$$

which is just one example of the general relation

$$\wp(\tau + \omega_3) \equiv e_3 + (e_2 - e_3) \operatorname{sn}^2(\kappa \tau \mid m). \tag{B.42}$$

Next, using the Jacobi elliptic solution (3.43) for $\epsilon > 2$, we find

$$z(\tau) = 2 \operatorname{sn}^2 \left(\sqrt{\epsilon/2}\,\tau \mid 2/\epsilon\right) = 2 \operatorname{sn}^2 \left(m^{1/2}\tau \mid m^{-1}\right), \tag{B.43}$$

and thus we recover once again the relation (B.42).

Appendix C

Noncanonical Hamiltonian Mechanics*

Modern formulations of Hamiltonian mechanics [1] rely on the use of non-canonical phase-space coordinates and the methods of differential geometry [5]. The purpose of this Appendix is to present a brief introduction to the noncanonical single-particle Hamiltonian mechanics. We also present an application of the (canonical) Hamiltonian perturbation method to the problem of the perturbed simple harmonic oscillator.

C.1   Differential Geometry

Differential k-forms

$$\omega_k = \frac{1}{k!} \omega_{i_1 i_2 \ldots i_k} \, dz^{i_1} \wedge dz^{i_2} \wedge \cdots \wedge dz^{i_k}$$

are fundamental objects in the differential geometry of n-dimensional space (with coordinates $\mathbf{z}$), where the components $\omega_{i_1 i_2 \ldots i_k}$ are antisymmetric with respect to interchange of two adjacent indices since the wedge product $\wedge$ is skew-symmetric (i.e., $dz^a \wedge dz^b = -dz^b \wedge dz^a$) with respect to the exterior derivative d (which has properties similar to the standard derivative d).

Note that the exterior derivative $d\omega_k$ of a differential k-form (or k-form for short) ω_k is a $(k+1)$-form. For example, the exterior derivative of a 0-form f is defined as

$$df \equiv \partial_a f \, dz^a, \tag{C.1}$$

and, thus, df is a differential 1-form; note that its components are the components of the gradient ∇f. Next, the exterior derivative of a 1-form Γ is a 2-form: $d\Gamma \equiv d\Gamma_b \wedge dz^b = \partial_a \Gamma_b \, dz^a \wedge dz^b$, which, as a result of the

245

skew-symmetry of the wedge product $\wedge$, may be expressed as

$$d\Gamma = \frac{1}{2}\left(\partial_a\Gamma_b - \partial_b\Gamma_a\right)dz^a \wedge dz^b$$

$$\equiv \frac{1}{2}\,\omega_{ab}\,dz^a \wedge dz^b, \tag{C.2}$$

where $\omega_{ab} = -\omega_{ba}$ denotes the antisymmetric components of the 2-form $\omega \equiv d\Gamma$.

An important difference between the exterior derivative d and the standard derivative d comes from the property that $d^2\omega_k = d(d\omega_k) \equiv 0$ for any k-form ω_k. Indeed, for a 0-form, we find

$$d^2 f = \partial_{ab}^2 f\,dz^a \wedge dz^b = 0,$$

since $\partial_{ab}^2 f$ is symmetric with respect to interchange $a \leftrightarrow b$ while $\wedge$ is antisymmetric. For a 1-form, we find

$$d^2\Gamma = \frac{1}{3!}\left(\partial_a\omega_{bc} + \partial_b\omega_{ca} + \partial_c\omega_{ab}\right)dz^a \wedge dz^b \wedge dz^c = 0,$$

which vanishes since $\partial_a\omega_{bc} + \partial_b\omega_{ca} + \partial_c\omega_{ab} \equiv 0$ vanishes identically since $\omega \equiv d\Gamma$.

A k-form ω_k is said to be *closed* if its exterior derivative is $d\omega_k \equiv 0$, while a k-form ω_k is said to be *exact* if it can be written in terms of a (k-1)-form Γ_{k-1} as $\omega_k \equiv d\Gamma_{k-1}$. Poincaré's Lemma states that all closed k-forms are exact (as can easily be verified), while its converse states that all exact k-forms are closed. For example, the infinitesimal volume element in three-dimensional space with curvilinear coordinates $\mathbf{u} = (u^1, u^2, u^3)$ and Jacobian $\mathcal{J}$: $\Omega \equiv \mathcal{J}(\mathbf{u})\,du^1 \wedge du^2 \wedge du^3$ is a closed 3-form since $d\Omega \equiv 0$. Hence, according to the converse of Poincaré's Lemma, there exists a 2-form σ such that $\Omega \equiv d\sigma$, where $\sigma \equiv \frac{1}{2}\,\epsilon_{ijk}\,\sigma^k(\mathbf{u})\,du^i \wedge du^j$ defines the infinitesimal area 2-form, with the Jacobian defined as $\mathcal{J} \equiv \partial\sigma^i(\mathbf{u})/\partial u^i$.

We now introduce the inner-product operation involving a vector field $\mathbf{v}$ and a k-form ω_k, denoted as $\mathbf{v} \cdot \omega_k$, which produces a $(k-1)$-form. For example, for a 1-form, it is defined as $\mathbf{v} \cdot \Gamma = v^a\,\Gamma_a$ while for a 2-form, it is defined as

$$\mathbf{v} \cdot \omega \equiv \frac{1}{2}\left(v^a\,\omega_{ab}\,dz^b - \omega_{ab}\,v^b\,dz^a\right) = v^a\,\omega_{ab}\,dz^b.$$

Note that $d(\mathbf{v} \cdot \Omega) = \mathcal{J}^{-1}\partial_a(\mathcal{J}\,v^a)\,\Omega \equiv (\nabla \cdot \mathbf{v})\,\Omega$, which can be used to derive the divergence of any vector field expressed in arbitrary curvilinear coordinates.

C.2   Lagrange and Poisson Tensors

The Poincaré-Cartan one-form [1] (Jules Henri Poincaré, 1854-1912; Élie-Joseph Cartan, 1869-1951) is expressed in canonical phase-space coordinates $(\mathbf{q}, \mathbf{p})$ as

$$\Gamma_c \equiv \frac{\partial L}{\partial \dot{q}^i}\, dq^i \, - \, H\, d\sigma \, = \, p_i\, dq^i \, - \, H\, d\sigma, \qquad \text{(C.3)}$$

where σ represents the Hamiltonian orbit parameter. The Lagrange one-form (a generalization of the Poincaré-Cartan one-form) is expressed in terms of general noncanonical phase-space coordinates z^a as

$$\Gamma \equiv \Lambda_a\, dz^a \, - \, H\, d\sigma, \qquad \text{(C.4)}$$

where

$$\Lambda_a \equiv \frac{\partial L}{\partial \dot{\mathbf{q}}} \cdot \frac{\partial \mathbf{q}}{\partial z^a}.$$

The two-form $\boldsymbol{\omega} \equiv d\gamma$ is written as

$$\boldsymbol{\omega} = \frac{1}{2}\, \omega_{ab}\, dz^a \wedge dz^b \, - \, dH \wedge d\sigma, \qquad \text{(C.5)}$$

where the components of the Lagrange two-form are

$$\omega_{ab} \equiv \frac{\partial \Lambda_b}{\partial z^a} \, - \, \frac{\partial \Lambda_a}{\partial z^b}. \qquad \text{(C.6)}$$

The phase-space Euler-Lagrange equation for the coordinate z^a is obtained by the contraction $\delta z^a \cdot \boldsymbol{\omega} \equiv 0$, which yields

$$\omega_{ab}\, \frac{dz^b}{d\sigma} \, = \, \frac{\partial H}{\partial z^a}. \qquad \text{(C.7)}$$

The noncanonical Hamilton's equations are obtained from Eq. (C.7) provided the antisymmetric Lagrange matrix $\boldsymbol{\omega}$ (with components ω_{ab}) can be inverted. This inversion condition is represented by $\det(\boldsymbol{\omega}) \neq 0$. The inverse of the Lagrange matrix yields the Poisson matrix $\mathsf{J} \equiv \boldsymbol{\omega}^{-1}$ with components J^{ab} that satisfy the condition $J^{ab}\, \omega_{bc} \equiv \delta^a_{\ c}$. The noncanonical Hamilton's equation (C.7) for z^a is therefore written as

$$\frac{dz^a}{d\sigma} = J^{ab}\, \frac{\partial H}{\partial z^b} \equiv \{z^a,\, H\}, \qquad \text{(C.8)}$$

where we introduced the antisymmetric Poisson bracket (Siméon-Denis Poisson, 1781-1840)

$$\{F,\, G\} \equiv \frac{\partial F}{\partial z^a}\, J^{ab}\, \frac{\partial G}{\partial z^b}. \qquad \text{(C.9)}$$

The antisymmetry of the Poisson matrix guarantees the antisymmetry of the Poisson bracket $\{G,\,F\} = -\,\{F,\,G\}$. An important property of the Poisson bracket is that it must satisfy the Jacobi condition (expressed in terms of three arbitrary functions F, G, and H)

$$\{F,\,\{G,\,H\}\} + \{G,\,\{H,\,F\}\} + \{H,\,\{F,\,G\}\} = 0, \qquad (\text{C.10})$$

which can be expressed in terms of the components of the Poisson matrix as

$$J^{ad}\,\partial_d J^{bc} + J^{bd}\,\partial_d J^{ca} + J^{cd}\,\partial_d J^{ab} = 0. \qquad (\text{C.11})$$

Note that when canonical coordinates are used (for which the Poisson components J^{ab} are either ± 1 or 0), the Jacobi identity (C.10) is trivially satisfied. The condition (C.11) can also be expressed in terms of the Lagrange components ω_{ab} as

$$\partial_a \omega_{bc} + \partial_b \omega_{ca} + \partial_c \omega_{ab} = 0,$$

which is trivially satisfied since the Lagrange components are defined by Eq. (C.6).

As a simple example of noncanonical Hamiltonian mechanics, we consider the Poincaré-Cartan one-form written in terms of the eight-dimensional noncanonical coordinates $z^a = (x^\mu,\,p_\mu) = (ct, \mathbf{x};\, w/c, \mathbf{p})$:

$$\Gamma = \left(p_\mu + \frac{e}{c}\,A_\mu\right)\,dx^\mu = \left(\mathbf{p} + \frac{e}{c}\,\mathbf{A}\right)\cdot d\mathbf{x} - (w - e\,\Phi)\,dt, \qquad (\text{C.12})$$

where the energy coordinate w is canonically conjugate to time t. The Lagrange two-form

$$\omega \equiv d\Gamma = dp_\mu \wedge dx^\mu + \frac{e}{2}\,F_{\mu\nu}\,dx^\mu \wedge dx^\nu \equiv \frac{1}{2}\,\omega_{ab}\,dz^a \wedge dz^b \qquad (\text{C.13})$$

is expressed in terms of the Faraday electromagnetic tensor components $F_{\mu\nu} \equiv \partial_\mu A_\nu - \partial_\nu A_\mu$. From the two-form (C.13), we construct the 8×8 antisymmetric Lagrange matrix

$$\boldsymbol{\omega} = \begin{pmatrix} (e/c)\,\mathsf{F} & -\mathbf{I} \\ \mathbf{I} & 0 \end{pmatrix},$$

which is composed of 4×4 block matrices. Its inversion yields the 8×8 antisymmetric Poisson matrix

$$\mathsf{J} \equiv \boldsymbol{\omega}^{-1} = \begin{pmatrix} 0 & \mathbf{I} \\ -\mathbf{I} & (e/c)\,\mathsf{F} \end{pmatrix},$$

from which we obtain the noncanonical Poisson bracket

$$\{F,\,G\} = \left(\frac{\partial F}{\partial x^\mu}\frac{\partial G}{\partial p_\mu} - \frac{\partial F}{\partial p_\mu}\frac{\partial G}{\partial x^\mu}\right) + \frac{e}{c}\,F^{\mu\nu}\frac{\partial F}{\partial p^\mu}\frac{\partial G}{\partial p_\nu}. \qquad (\text{C.14})$$

When we combine this Poisson bracket with the Hamiltonian $H = p_\mu p^\mu / 2m$, we obtain Hamilton's equations of motion

$$z^a = \{z^a, H\} = \begin{cases} \dot{x}^\mu = v^\mu \\ \\ \dot{p}_\mu = (e/c) F_{\mu\nu} v^\nu \end{cases}$$

which describe the relativistic motion of a particle in an electromagnetic field [7].

Lastly, an important property of Hamilton's equations is that the equations satisfy the Liouville Theorem

$$\frac{\partial}{\partial z^a} \left(\mathcal{J} \frac{dz^a}{d\sigma} \right) = 0, \tag{C.15}$$

which can also be expressed in terms of the Liouville identities $\partial_a (\mathcal{J} J^{ab}) = 0$, where the Jacobian $\mathcal{J}$ is the determinant of the matrix $\partial(\mathbf{q}, \mathbf{p})/\partial\mathbf{z}$. For the Poisson bracket (C.14), the Liouville identities are $\partial F_{\mu\nu}/\partial p_\mu \equiv 0$. Lastly, the Liouville Theorem is trivially satisfied in the case of canonical Hamilton's equations.

C.3   Hamiltonian Perturbation Theory

Hamiltonian methods offer powerful tools in perturbation theory. For example, when an exact dynamical invariant is destroyed by a small perturbation, Hamiltonian perturbation methods can be used to construct an adiabatic invariant that is preserved to arbitrary order in the perturbation amplitude.

We consider, for example, the unperturbed (canonical) Hamiltonian $H_0 = p^2/2 + q^2/2$ for a simple harmonic oscillator with unit mass and unit frequency. By introducing the transformation to *action-angle* coordinates $\mathbf{z} = (J, \theta)$, where $q = \sqrt{2J} \sin\theta$ and $p = \sqrt{2J} \cos\theta$, we readily find that the new unperturbed Hamiltonian $K_0(\mathbf{z}) \equiv H_0(q(\mathbf{z}), p(\mathbf{z})) = J$ is independent of the angle θ. Hence, the new unperturbed Hamilton's equations are $\dot{J}_0 = -\partial K_0/\partial\theta \equiv 0$ and $\dot{\theta}_0 = \partial K_0/\partial J \equiv 1$. The action variable is therefore an invariant of the unperturbed Hamiltonian system.

We now introduce the perturbation Hamiltonian $\epsilon H_1(q, p) = -\epsilon q^4/24$ in the original simple-harmonic-oscillator Hamiltonian system (where ϵ appears as an ordering parameter), which is translated into the new perturbation Hamiltonian

$$\epsilon K_1(J, \theta) = -\frac{\epsilon}{6} J^2 \sin^4\theta. \tag{C.16}$$

We note here that in order for the perturbation to be considered small (i.e., $|K_1| < K_0$), we require that $\epsilon < 6/J_{max}$ during the evolution of the perturbed system. Because the new Hamiltonian $K \equiv K_0 + \epsilon K_1$ now depends on the angle variable θ, the action variable J is no longer invariant, i.e., $\dot{J} = -\epsilon \, \partial K_1/\partial\theta \neq 0$. The purpose of Hamiltonian perturbation theory is to construct new action-angle coordinates $\overline{z} = (\overline{J}, \overline{\theta})$ in terms of which the transformed Hamiltonian

$$\overline{K}(\overline{z}) \;=\; K(J(\overline{z}), \theta(\overline{z})) \tag{C.17}$$

is independent of the new angle variable $\overline{\theta}$ up to arbitrary orders in ϵ.

Since the transformation $(J, \theta) \to (\overline{J}, \overline{\theta})$ we seek is canonical, it may be expressed in the following form

$$\overline{z}^a \;=\; z^a + \epsilon\{S_1, \, z^a\} + \epsilon^2 \left(\{S_2, \, z^a\} + \frac{1}{2}\left\{S_1, \, \{S_1, \, z^a\}\right\} \right) + \cdots, \tag{C.18}$$

where the functions $(S_1, S_2, \cdots)$ are said to *generate* the canonical transformation, and the action-angle canonical Poisson bracket is $\{F, G\} = \partial_\theta F \, \partial_J G - \partial_J F \, \partial_\theta G$. The new Hamiltonian, on the other hand, is expressed in terms of these generating functions as

$$\overline{K} \;=\; K - \epsilon\{S_1, \, K\} - \epsilon^2 \left(\{S_2, \, K\} + \frac{1}{2}\left\{S_1, \, \{S_1, \, K\}\right\} \right) + \cdots, \tag{C.19}$$

which ensures the scalar-invariance property (C.17) is satisfied. Note that the *direct* transformation approach used here is different from the standard perturbation analysis based on mixed-variable generating functions [7]. The main advantage of the direct approach is that it can easily be generalized to arbitrary orders in the perturbation parameter ϵ.

When the original Hamiltonian

$$K \;=\; K_0(J) + \epsilon K_1(J, \theta) + \epsilon^2 K_2(J, \theta) + \cdots \tag{C.20}$$

is expanded in powers of ϵ with each perturbation term $K_n(J, \theta)$ $(n \geq 1)$ expressed as an explicit function of θ, the transformed Hamiltonian (C.19) is also expressed as an expansion in powers of ϵ

$$\overline{K} \;=\; \overline{K}_0(\overline{J}) + \epsilon \overline{K}_1(\overline{J}) + \epsilon^2 \overline{K}_2(\overline{J}) + \cdots \tag{C.21}$$

By inserting Eqs. (C.20)-(C.21) in Eq. (C.19), we obtain the following expressions up to second order in ϵ:

$$\overline{K}_0 = K_0, \tag{C.22}$$

$$\overline{K}_1 = K_1 - \{S_1, K_0\} = K_1 - \frac{\partial S_1}{\partial \theta}, \tag{C.23}$$

$$\overline{K}_2 = K_2 - \{S_2, K_0\} - \{S_1, K_1\} + \frac{1}{2}\left\{S_1, \{S_1, K_0\}\right\}$$

$$= K_2 - \frac{\partial S_2}{\partial \theta} - \overline{K}_1' \frac{\partial S_1}{\partial \theta} - \frac{1}{2}\left\{S_1, \frac{\partial S_1}{\partial \theta}\right\}. \tag{C.24}$$

At zeroth order, we easily find $\overline{K}_0 = \overline{J}$. At first order, we impose the condition that $\overline{K}_1$ is independent of $\overline{\theta}$ by $\overline{\theta}$-averaging the right side of Eq. (C.23), which yields

$$\overline{K}_1(\overline{J}) \equiv \langle K_1(\overline{J}, \overline{\theta})\rangle, \tag{C.25}$$

while the $\overline{\theta}$-dependent part $\widetilde{K}_1 \equiv K_1 - \langle K_1\rangle$ is cancelled by choosing S_1 such that $\partial S_1/\partial\overline{\theta} \equiv \widetilde{K}_1$. Likewise, the transformed second-order Hamiltonian is defined as

$$\overline{K}_2 \equiv \langle K_2\rangle - \frac{1}{2}\frac{\partial}{\partial\overline{J}}\left\langle\left(\frac{\partial S_1}{\partial\theta}\right)^2\right\rangle = \langle K_2\rangle - \frac{1}{2}\frac{\partial\langle(\widetilde{K}_1)^2\rangle}{\partial\overline{J}}, \tag{C.26}$$

while the second-order generating function S_2 is chosen to cancel all explicit $\overline{\theta}$-dependence on the right side of Eq. (C.24).

We note that since the new action variable

$$\overline{J} = J + \epsilon\frac{\partial S_1}{\partial\theta} + \epsilon^2\left(\frac{\partial S_2}{\partial\theta} + \frac{1}{2}\left\{S_1, \frac{\partial S_1}{\partial\theta}\right\}\right) + \cdots \tag{C.27}$$

is expressed in terms of a truncated asymptotic series in powers of ϵ, it is not an exact dynamical invariant, i.e., $\dot{\overline{J}} = \mathcal{O}(\epsilon^{n+1})$ if the new Hamiltonian $\overline{K} = \overline{K}_0 + \cdots + \epsilon^n\overline{K}_n$ is truncated at order ϵ^n. Hence, the new action variable (C.27) called an *adiabatic* invariant [12].

Returning to the perturbation term (C.16), for example, we find

$$\langle K_1\rangle = -\frac{J^2}{16} \quad\text{and}\quad \widetilde{K}_1 = -\frac{J^2}{48}(\cos 4\theta - 4\cos 2\theta) \equiv \frac{\partial S_1}{\partial\theta},$$

so that, up to first order in ϵ, the new Hamiltonian is

$$\overline{K} = \overline{J} - \frac{\epsilon}{16}\overline{J}^2, \tag{C.28}$$

while the new action-angle variables are

$$\overline{J} = J - \frac{\epsilon}{48}J^2(\cos 4\theta - 4\cos 2\theta) \quad\text{and}\quad \overline{\theta} = \theta + \frac{\epsilon}{96}J(\sin 4\theta - 8\sin 2\theta).$$

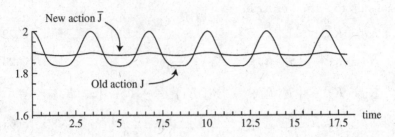

Fig. C.1   Plots of the old action $J(t)$ and the new action $\overline{J}(t)$ (with first-order corrections) corresponding to the initial conditions $q(0) = 2$ and $p(0) = 0$ with perturbation parameter $\epsilon = 0.25$.

If we now write $\overline{J}$ in terms of the original coordinates (q, p), we find

$$\overline{J}' = \frac{1}{2}\left(p^2 + q^2\right) - \frac{\epsilon}{192}\left(5\,q^4 - 6\,p^2 q^2 - 3\,p^4\right), \qquad (C.29)$$

and we easily verify that $\dot{\overline{J}} = \mathcal{O}(\epsilon^2)$, with $\dot{q} = p$ and $\dot{p} = -q + \epsilon\,q^3/6$.

Figure C.1 shows plots of the old action $J(t) = p^2(t)/2 + q^2(t)/2$ and the new action $\overline{J}(t)$, given by Eq. (C.29), as functions of time t for $\epsilon = 0.25$. Note that since $1 < J \leq 2$ during its time evolution, then $\epsilon = 0.25 < 6/J_{\max} = 3$ satisfies the condition of applicability of Hamiltonian perturbation theory. One can clearly see that, even for a large value of the perturbation parameter ϵ, the new action $\overline{J}$ shows much smaller oscillations than the old action J. One could further reduce the oscillations in the new action $\overline{J}$ by proceeding to second order in the Hamiltonian perturbation analysis [see Eq. (C.27)], which requires us to evaluate the generating function S_2 in Eq. (C.24).

Lastly, we note that the simplicity of the new Hamilton's equations of motion

$$\dot{\overline{J}} = -\frac{\partial \overline{K}}{\partial\overline{\theta}} = 0 \quad \text{and} \quad \dot{\overline{\theta}} = \frac{\partial \overline{K}}{\partial\overline{J}} = 1 - \frac{\epsilon}{8}\overline{J} \equiv \overline{\Omega}, \qquad (C.30)$$

implies that the old action-angle variables can be evaluated explicitly as functions of time by inverting the transformation (C.18):

$$J(t) = \overline{J} + \frac{\epsilon}{48}\overline{J}^2\left[\cos 4(\overline{\theta}_0 + \overline{\Omega}t) - 4\cos 2(\overline{\theta}_0 + \overline{\Omega}t)\right], \qquad (C.31)$$

$$\theta(t) = \overline{\theta}_0 + \overline{\Omega}t - \frac{\epsilon}{96}\overline{J}\left[\sin 4(\overline{\theta}_0 + \overline{\Omega}t) - 8\sin 2(\overline{\theta}_0 + \overline{\Omega}t)\right]. \qquad (C.32)$$

By extension, the old coordinates

$$q(t) = \sqrt{2J(t)}\,\sin\theta(t) \quad \text{and} \quad p(t) = \sqrt{2J(t)}\,\cos\theta(t) \qquad (C.33)$$

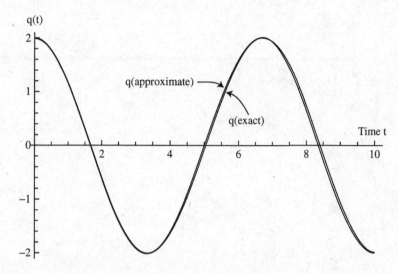

Fig. C.2 Plots of exact solution $q(t)$ of the perturbed Hamiltonian problem (C.34) and the approximate solution $q(t) = \sqrt{2J(t)} \sin \theta(t)$, with $(J(t), \theta(t))$ given by Eqs. (C.31)-(C.32), for the same initial condition $(q(0), p(0)) = (2, 0)$ and $\epsilon = 0.25$.

have also been solved explicitly as functions of time. Hamiltonian perturbation theory has therefore allowed us to solve explicitly the Hamilton's equations

$$\dot{q} = p \quad \text{and} \quad \dot{p} = -q + \frac{\epsilon}{6} q^3 \tag{C.34}$$

for small enough values of the perturbation parameter ϵ. It is important to note that the solution (C.33), with $(J(t), \theta(t))$ given by Eqs. (C.31)-(C.32), starts to deviate from the true solution of Eq. (C.34) for times of order ϵ^{-n} when the Hamiltonian perturbation analysis has been carried out up to order ϵ^n. For example, Fig. C.2 shows that the approximate solution (C.33) begins to deviate from the exact solution at a time close to $\epsilon^{-1} = 4$. Note that the deviation of the approximate solution oscillates around the exact solution and the amplitude of the deviation depends on the initial conditions (i.e., how well the perturbation condition $\epsilon < 6/J_{\max}$ is satisfied).

Bibliography

[1] Arnold, V. I., *Mathematical Methods of Classical Mechanics, 2nd ed.* (Springer-Verlag, 1989).

[2] Basdevant, J.-L., *Variational principles in physics* (Springer, 2006).

[3] Born, M., and Wolf, E., *Principles of Optics, 5th ed.* (Pergamon Press, 1975).

[4] Dugas, R., *A History of Mechanics* (Dover, 1988).

[5] Flanders, H., *Differential Forms with Applications to the Physical Sciences* (Dover, 1989).

[6] Fox, C., *An Introduction to the Calculus of Variations* (Dover, 1987).

[7] Goldstein, H., Poole, C., and Safko, J., *Classical Mechanics, 3rd ed.* (Addison Wesley, 2002).

[8] Gray, C. G., Karl, G., and Novikov, V. A., *The four variational principles of mechanics*, Ann. Phys. **251**, 1-25 (1996).

[9] Gray, C. G., Karl, G., and Novikov, V. A., *From Maupertuis to Schroedinger: Quantization of classical variational principles*, Am. J. Phys. **67**, 959-961 (1999).

[10] Gray, C. G., Karl, G., and Novikov, V. A., *Progress in classical and quantum variational principles*, Rep. Prog. Phys. **67**, 159-208 (2004).

[11] Lanczos, C., *The Variational Principles of Mechanics, 4th ed.* (Dover, 1970).

[12] Landau, L. D., and Lifshitz, E. M., *Mechanics, 3rd ed.* (Pergamon Press, 1976).

[13] Marion, J. B., and Thornton, S. T., *Classical Dynamics of Particles and Systems, 4th, ed.* (Harcourt, 1995).

[14] Milne-Thomson, L. M., *Jacobi Elliptic Functions and Theta Functions* in *Handbook of Mathematical Functions*, Abramowitz, M., and Stegun, I. A., eds. (Dover, New York, 1965) chap. 16.

[15] Nesbet, R. K., *Variational principles and methods in theoretical physics and chemistry* (Cambridge University Press, 2003).

[16] Southard, T. H., *Weierstrass Elliptic and Related Functions* in *Handbook of Mathematical Functions*, Abramowitz, M., and Stegun, I. A., eds. (Dover, New York, 1965) chap. 18.

[17] Whittaker, E. T., *A Treatise on the Analytical Dynamics of Particles and Rigid Bodies, 4th ed.* (Dover, 1944).

[18] Yourgrau, W., and Mandelstam, S., *Variational Principles in Dynamics and Quantum Theory* (Dover, 1968).

Index